ASTONISHING PREHISTORIC KNOWLEDGE

PROFESSOR DOCTOR SANJAY ROUT

Astonishing Prehistoric Knowledge written by Professor Sanjay Rout and Edited by Professor Prangyan Biswal Published by ISL Publications, India

© Author

INNOVATION SOLUTION LAB
DREAM. CREATE. INSPIRE.

ISL Publications

India

Website- https://innovationsolutionlab.weebly.com

Contents

Astonishing Prehistoric Knowledge

-Author-
Professor Doctor Sanjay Rout

ISL Publications

Foreword

This Book is Been forwarded to all Global Innovators & Researchers , God and Parents

Preface

The book is published under innovation solution lab media & publication Going back a large number of years are various instances of antiquated innovation that leave us amazed at the information and intelligence held by individuals of our past. The book amazingly enlighten all about how ancient technologies are impactful and if it can use in certain way , how impactful it will change the civilization.

Acknowledgements

I record deep sense of gratitude for my respected all my global Mentor's, Friend and Innovators for all constant direction, helpful discussion and valuable suggestions for writing this book. Due to his valuable suggestions and regular encouragement. I would be able to complete this work and fulfillment of my dream. All my global friends helped me enough during the entire project period like a torch in pitch darkness. I shall remain highly indebted to all throughout my life.

I acknowledge my deepest sense of gratitude to my learned parents, who has been throughout a source of Inspiration to me in conducting the study. Who helped me at various stages of the study directly or indirectly. He also enlightened me to follow the path of duty.

Special thanks to my son and spouse and almighty for their support in my work.

Prologue

Introduction

Going back a large number of years are various instances of antiquated innovation that leave us dazed at the information and knowledge held by individuals of our past. They were the aftereffect of unimaginable advances in building and development as new, amazing human advancements rose and came to rule the antiquated world. These advances animated social orders to receive better approaches for living and administration, just as better approaches for understanding their reality. Be that as it may, numerous old innovation puzzles were overlooked, lost to the pages of history, just to be re-concocted centuries later. Here we highlight antiquated innovation history and many stunning relics that mirror the brightness of old personalities.

We've lost the key to making a portion of history's most helpful innovations, and for the entirety of our resourcefulness and revelations, our progenitors of thousands of years prior are as yet ready to bewilder us with their creativity and disclosures. We have built up the cutting edge likeness a portion of these developments, yet without a doubt, as of late.

1. Greek Fire: Mysterious Chemical Weapon

The Byzantines of the seventh to twelfth hundreds of years heaved a baffling substance at their foes in maritime fight. This fluid, shot through cylinders or siphons, consumed in water and must be stifled with vinegar, sand, and pee. We despite everything don't have the foggiest idea what this compound weapon , known as Greek Fire, was made of. The Byzantines protected the mystery enviously, guaranteeing just a chosen few knew the mystery, and the information was in the end lost through and through .

2. Adaptable Glass: A Substance Too Precious

Three old records of a substance known as vitrum flexile , adaptable glass, are not satisfactory enough to verify that this substance really existed. The account of its innovation was first told by Petronius (d. 63 A.D.).

He expounded on a glassmaker who introduced the Emperor Tiberius (who ruled 14–37 A.D.) with a glass vessel. He requested that the ruler hand it back to him, so, all in all, the glassmaker tossed it to the floor. It didn't break; it just scratched, and the glassmaker pounded it rapidly once again into shape. Dreading the depreciation of valuable metals, Tiberius requested the creator guillotined so the mystery of vitrum flexile would kick the bucket with him.

Pliny the Elder (d. 79 A.D.) recounted to this story too. He said that, in spite of the fact that the story was much of the time advised, it may not be altogether obvious.

The adaptation told several hundred years after the fact by Dio Cassius transformed the glassmaker into such an entertainer. At the point when the vessel was tossed to the floor, it broke and the glassmaker fixed it with his exposed hands.

In 2012, the glass fabricating organization Corning presented its adaptable "Willow Glass." Heat-safe and adaptable enough to be moved up, it has demonstrated particularly valuable in making sun based boards.

On the off chance that the disastrous Roman glassmaker did without a doubt design vitrum flexile, it appears he was a large number of years relatively revolutionary.

3. An Antidote to All Poisons

A purported "all inclusive antitoxin" against all toxic substances was said to have been created by King Mithridates VI of Pontus (who ruled 120–63 B.C.) and idealized by Emperor Nero's own doctor. The first equation was lost, clarified Adrienne Mayor, a folklorist and antiquarian of science at Stanford University, in a 2008 paper, named "Greek Fire, Poison Arrows and Scorpion Bombs: Biological and Chemical Warfare in the Ancient World." But old students of history disclosed to us that among its fixings were opium, hacked snakes, and a blend of little dosages of toxic substances and their cures.

The important substance was known as Mithridatium, named for King Mithridates VI.

City hall leader noticed that Serguei Popov, a previous top organic weapons scientist in the Soviet Union's monstrous Biopreparat program who abandoned to the United States in 1992, was endeavoring to make a present day Mithridatium.

4. Warmth Ray Weapon

Greek mathematician Archimedes (d. 212 B.C.) built up a warmth beam weapon that resisted the abilities of Discovery Channel's "Mythbusters" to recreate in 2004. City hall leader portrayed the weapon as "positions of cleaned bronze shields mirroring the sun's beams at foe ships."

In spite of the fact that "Mythbusters" neglected to repeat this old weapon and proclaimed it a fantasy, MIT understudies prevailing in 2005. They combusted a pontoon in San Francisco harbor utilizing the 2,200-year-old weapon.

A warmth beam weapon divulged in 2001 by the Defense Advanced Research Projects Agency (DARPA) utilized microwaves to enter "a casualty's skin, warming it to 130 degrees Fahrenheit, making the vibe that one is ablaze," clarified Mayor.

5. Roman Concrete

The tremendous Roman structures that have kept going a huge number of years are demonstrations of the points of interest Roman cement has over the solid utilized these days, which gives indications of debasement following 50 years.

Specialists have worked as of late to reveal the mystery of this antiquated solid's life span. The mystery fixing is volcanic debris.

An article distributed in 2013 by the University of California–Berkeley News Center reported that college analysts portrayed just because how the exceptionally steady compound calcium-aluminum-silicate-hydrate (C-A-S-H) ties the material. The way toward making it would make lower carbon dioxide outflows than the procedure for making current cement. A few weaknesses of its utilization are, notwithstanding, that it takes more time to dry, and despite the fact that it keeps going longer, it is more vulnerable than present day concrete.

Chapter-I
Ancient Technologies

space-886060_640.jpg

Could Ancient Peruvians Soften Stone?

Sacsayhuamán is a stronghold on the northern edges of the city of Cusco, Peru, the noteworthy capital of the Inca Empire.

Archeologists and different researchers have been scratching their heads attempting to make sense of how wonderful old Peruvian structures like Sacsayhuamán were developed. This glorious structure comprises of colossal stones so substantial that our cutting edge hardware can scarcely move and set up. Does the way in to the riddle lie in a specific unmistakable plant that gave the old Peruvians plausibility to mellow stone or is the response to the secret access to cutting edge old innovation that could soften stone?

As indicated by scientists Jan Peter de Jong , Christopher Jordan Jesus Gamarra, the stone dividers in Cuzco show proof of being warmed to an extremely serious extent and vitrified-the outside surface turning out to be polished and exceptionally smooth. In light of this perception, Jong, Jordan and Gamarra make the determination that "an innovative gadget was utilized to dissolve stone squares which were then positioned and permitted to cool close to hard, jigsaw-polygonal obstructs that were at that point set up.

Analysts Jong and Jordan recommend that few antiquated human advancements overall knew about the cutting edge softening of stone innovation. They likewise state that "the stones on a portion of the old avenues in Cuzco have been vitrified by some high temperature to give them their trademark smooth surface

2. Incamisana Water Temple At Ollantaytambo, Peru: Marvelous Engineering Masterpiece Of Inca

The Incas were known for their designing practices, especially as to water building. Their plan of water frameworks was impacted by their solid convictions in intensity of water.

They considered water to be both a physical and profound wonder. They considered their extraordinary human progress as something that emerges from water.

The Incamisana sanctuary at Ollantaytambo in Peru was built for love of divinities who gave the Inca individuals water and water itself. Indeed, even today in our cutting edge terms, the structure speaks to the Inca's perfect work of art of both structural designing and development. It was intended to concentrate on strict services and contributions.

It's certainly a really momentous accomplishment particularly, in the event that we think about the district's unforgiving climate and the height of 2,792 m (9,160 ft) above ocean level.

Flooding that occurred in 1679 harmed the sanctuary, yet the store of residue luckily saved the lower bit of it.

3. The Lycurgus Cup: Fascinating Artifact That Reveals Prehistoric Knowledge Of Nanotechnology

This astonishing antiquity plainly shows that our predecessors were a long ways comparatively radical. Truth be told they were progressed to the point that they may even have been creators of what we today call nanotechnology. This remarkable cup is the main enduring total model produced using dichroic glass, which changes shading when held up to the light. At the point when light is shone through the body of the cup it abandons obscure green to a gleaming translucent red. The glass contains small measures of colloidal gold and silver, which give it these uncommon optical properties.

Another genuine model uncovering our progenitors knew about nanotechnology are the bizarre infinitesimally little nanospirals inside material that must be at any rate 100,000 years of age have been distinguished in a few spots, during a normal examination of mineral stores in the Ural Mountains in 1992. The starting point of these uncommon however weird curios that the natural eye can scarcely observe - is still not clarified.

4. Batteries In The Ancient World

A little, undecorated ancient rarity with rather plain appearance, is accepted by certain researchers to be a case of an ancient, electrical force source. It's the alleged Baghdad Battery, otherwise called the Parthian Battery. The relic - thought to be a 2,000-year-old electric battery - was found in 1936 by railroad laborers in the zone of Tel Khujut Rabu, south of Baghdad. Most sources date the batteries to around 200 BC, however the main known electric battery - the Voltaic heap - was not created by Italian physicist Alessandro Volta until 1799.

The bottoms of these puzzling chambers were topped with copper plates and fixed with bitumen or black-top. Another protecting layer of bitumen fixed the highest points of the pots and held set up iron poles suspended into the focal point of the copper chambers. The poles demonstrated a solid proof of having been eroded by a corrosive arrangement that dissipated quite a while in the past.

5. Mind blowing Ancient Metallurgical Wonders

The section in India - comprised of 98% created iron of sullied quality - not at all welded together - appears to have been manufactured as a solitary, enormous bit of iron.

The Antediluvians had innovations that coordinated our own; there are additionally genuine signs that in specific regions they even had remarkable information, which has just scarcely been poked by our present-day science. Profoundly propelled solidifying methods of the people of yore just as old castings of huge pieces were far reaching in ancient times. Our progenitors were in control of an incredibly complex logical information on metalworking from a previous human progress and proof of this information was found in various pieces of the world. China with a long history in metallurgy, was the soonest progress that fabricated cast iron and a portion of the old Chinese accomplishments of throwing iron are so amazing as to be practically unimaginable.

Old Indians, for instance, created iron equipped for withstanding erosion, in all probability because of the high phosphorus substance of the iron delivered during those occasions. A segment of cast iron 23 feet (7 meters) high, weighing roughly 6 tons with distance across of 16.4 inches remains in the yard of Kutb Minar in Delhi, India.

An engraving in the Sanskrit language illuminates that the segment was initially raised in the sanctuary of Muttra and topped with Garuda - "Courier of the Gods" - a picture of the winged creature manifestation of the god Vishnu, the Indian god known as "The Preserver".

6. Water systems Are Among Most Exceptional Achievements Of Ancient Roman Engineers

In 312 BC, a Roman lawmaker Appius Claudius Caecus requested that a reservoir conduit be worked to gracefully Rome with water. The water passage was one of two significant Roman tasks did during this period; the other was a street for military purposes.

Among all the accomplishments of old designing, Roman water systems have a place with the most uncommon ones.

Specialists in old Roman Empire realized that one of the most central prerequisites for any town or city to exist is to flexibly it with water, since it is essential human need.

The city must have water to drink, cook and clean, yet it likewise should have every single crucial course of action to expel undesirable water. Both old and current Rome has been very much provided with the methods for conveying water to the city and removing it.

Antiquated Romans were exceptionally gifted specialists. By the mid fourth century AD, Rome was being provided by in excess of twelve reservoir conduits, which helpfully could bring in excess of a million cubic meters of new water to the city consistently! This gigantic measure of water was conveyed to Rome's occupants through a mind boggling system of tanks and pipes and to around 1,500 open wellsprings, pools and right around 900 open and private showers.

7. Amazing Viking Crystal Sunstones Helped Vikings To Navigate

Here and there, the innovation was utilized to apply genuine gold and silver. It additionally was utilized falsely, to make modest metal sculptures that resemble strong gold or silver.

People of old were in control of modern information. Old gold and silversmiths utilized mercury, which was created over 8,000 years back in Turkey. Mercury was utilized for plating (vaults, insides of houses of God, strict figures and that's only the tip of the iceberg) in numerous pieces of the old world. A considerable lot of the people of yore's procedures are as yet obscure. They were gifted to such an extent that a portion of the quality they accomplished has still not been coordinated. In antiquated occasions, these advanced techniques were utilized to create and brighten various sorts of relics, for example, gems, sculptures, special necklaces, and ordinarily utilized articles.

Once in a while, the innovation was utilized to apply genuine gold and silver. It likewise was utilized deceitfully, to make modest metal sculptures that resemble strong gold or silver.

People of old were in control of refined information. Antiquated gold and silversmiths utilized mercury, which was delivered over 8,000 years prior in Turkey. Mercury was utilized for plating (arches, insides of houses of prayer, strict figures and the sky is the limit from there) in numerous pieces of the antiquated world. A large number of the people of yore's procedures are as yet obscure. They were talented to the point that a portion of the quality they accomplished has still not been coordinated. In antiquated occasions, these refined techniques were utilized to create and enliven various sorts of ancient rarities, for example, gems, sculptures, talismans, and ordinarily utilized items.

Gilders played out these procedures not exclusively to finish protests yet in addition to mimic the presence of gold or silver, once in a while falsely. From an innovative perspective, the point of these laborers more than 2000 years back was to make the valuable metal coatings as slim and follower as could be expected under the circumstances.

This was so as to spare costly metals and to improve the protection from the wear brought about by proceeded with use and course. Late discoveries affirm the significant level of capability came to by the old specialists and skilled workers and stresses a creative nature of the articles they delivered couldn't be bettered in old occasions and has not yet been reached in present day ones. Understand more

9. Trend setting innovation Of The Ancients: Artificial Platforms Of Mighty Nan Madol

Two truly exceptional islands - Kosrae and Pohnpei ("upon a stone special stepped area") - can be found in the Caroline Islands, a generally dissipated archipelago situated in the western Pacific Ocean, toward the north of New Guinea and in excess of 1,500 miles toward the east of the Philippines.

These two spots are loaded with puzzles, especially in view of strange survives from old design, in type of old gigantic urban communities, of which starting point nobody can figure."

Those, who don't refute the fantasies and legends of the people of yore, can consider over some intriguing legends and legends. One of them says that "enchantment powers" were utilized to ship the logs from a distant area, by making them fly but then another legend underpins the last mentioned, saying about a compelling "entertainer" who made the logs fly towards Nan Madol, where the city was developed and a firebreathing mythical serpent turned into its image.

Who was keen on building two of the Pacific's biggest basalt urban communities? Understand more

10. Exceptionally Advanced Robots In Ancient China

A portion of these refined old developments were later re-designed by our advanced society, yet not all. There is no uncertainty antiquated individuals had profoundly propelled information in various regions.

1. Could Ancient Peruvians Soften Stone?

Sacsayhuaman, Peru

Sacsayhuamán is a bastion on the northern edges of the city of Cusco, Peru, the memorable capital of the Inca Empire.

Archeologists and different researchers have been scratching their heads attempting to make sense of how wonderful old Peruvian structures like Sacsayhuamán were built. This heavenly structure comprises of monstrous stones so overwhelming that our cutting edge apparatus can barely move and set up. Does the way in to the riddle lie in a specific quite certain plant that gave the antiquated Peruvians probability to relax stone or is the response to the secret access to cutting edge old innovation that could liquefy stone?

As per scientists Jan Peter de Jong , Christopher Jordan Jesus Gamarra, the stone dividers in Cuzco show proof of being warmed to an exceptionally serious extent and vitrified-the outside surface turning out to be shiny and smooth. In light of this perception, Jong, Jordan and Gamarra reach the determination that "an innovative gadget was utilized to dissolve stone squares which were then positioned and permitted to cool close to hard, jigsaw-polygonal obstructs that were at that point set up.

Scientists Jong and Jordan recommend that few antiquated human advancements overall knew about the cutting edge softening of stone innovation. They additionally state that "the stones on a portion of the antiquated roads in Cuzco have been vitrified by some high temperature to give them their trademark smooth surface.

2. Incamisana Water Temple At Ollantaytambo, Peru: Marvelous Engineering Masterpiece Of Inca

Incamisana Water Temple At Ollantaytambo, Peru: Marvelous Engineering Masterpiece Of Inca

One of numerous wellsprings of the Incamisana Water Temple.

The Incas were known for their designing practices, especially with respect to water building. Their plan of water frameworks was impacted by their solid convictions in intensity of water.

They considered water to be both a physical and profound wonder. They considered their incredible human progress as something that emerges from water.

The Incamisana sanctuary at Ollantaytambo in Peru was built for love of divinities who gave the Inca individuals water and water itself. Indeed, even today in our cutting edge terms, the structure speaks to the Inca's gem of both structural building and development. It was intended to concentrate on strict services and contributions.

It's certainly a really wonderful accomplishment particularly, on the off chance that we think about the district's unforgiving climate and the elevation of 2,792 m (9,160 ft) above ocean level.

Flooding that occurred in 1679 harmed the sanctuary, yet the store of dregs luckily safeguarded the lower bit of it.

3. The Lycurgus Cup: Fascinating Artifact That Reveals Prehistoric Knowledge Of Nanotechnology

The Lycurgus Cup: Fascinating Artifact That Reveals Prehistoric Knowledge Of Nanotechnology

The Lycurgus cup.

This stunning ancient rarity obviously shows that our progenitors were a long ways comparatively radical. Indeed they were progressed to such an extent that they may even have been innovators of what we today call nanotechnology. This uncommon cup is the main enduring total model produced using dichroic glass, which changes shading when held up to the light. At the point when light is shone through the body of the cup it abandons obscure green to a shining translucent red. The glass contains little measures of colloidal gold and silver, which give it these irregular optical properties. Understand more

Another genuine model uncovering our predecessors knew about nanotechnology are the odd minutely little nanospirals inside material that must be at any rate 100,000 years of age have been identified in a few spots, during a normal examination of mineral stores in the Ural Mountains in 1992. The inception of these remarkable however bizarre ancient rarities that the natural eye can scarcely observe - is still not clarified.

4. Batteries In The Ancient World

Antiquated Baghdad batteries

A little, undecorated antiquity with rather plain appearance, is accepted by certain researchers to be a case of an ancient, electrical force source. It's the alleged Baghdad Battery, otherwise called the Parthian Battery. The ancient rarity - thought to be a 2,000-year-old electric battery - was found in 1936 by railroad laborers in the territory of Tel Khujut Rabu, south of Baghdad. Most sources date the batteries to around 200 BC, however the main known electric battery - the Voltaic heap - was not concocted by Italian physicist Alessandro Volta until 1799.

The bottoms of these puzzling chambers were topped with copper plates and fixed with bitumen or black-top. Another protecting layer of bitumen fixed the highest points of the pots and held set up iron poles suspended into the focal point of the copper chambers. The poles demonstrated a solid proof of having been eroded by a corrosive arrangement that dissipated quite a while in the past.

5. Unbelievable Ancient Metallurgical Wonders

The section in India - comprised of 98% fashioned iron of tainted quality - not at all welded together - appears to have been produced as a solitary, immense bit of iron.

The section in India - comprised of 98% fashioned iron of tainted quality - not at all welded together - appears to have been produced as a solitary, immense bit of iron.

The Antediluvians had innovations that coordinated our own; there are additionally genuine signs that in specific territories they even had exceptional information, which has just barely been bumped by our present-day science. Profoundly propelled solidifying strategies of the people of old just as old castings of enormous pieces were boundless in days of yore. Our progenitors were in control of an incredibly advanced logical information on metalworking from a previous human progress and proof of this information was found in various pieces of the world. China with a long history in metallurgy, was the soonest development that fabricated cast iron and a portion of the old Chinese accomplishments of throwing iron are so noteworthy as to be practically extraordinary.

Old Indians, for instance, created iron equipped for withstanding erosion, no doubt because of the high phosphorus substance of the iron delivered during those occasions. A section of cast iron 23 feet (7 meters) high, weighing around 6 tons with distance across of 16.4 inches remains in the patio of Kutb Minar in Delhi, India.

An engraving in the Sanskrit language educates that the segment was initially raised in the sanctuary of Muttra and topped with Garuda - "Delivery person of the Gods" - a picture of the fowl manifestation of the god Vishnu, the Indian god known as "The Preserver".

6. Reservoir conduits Are Among Most Exceptional Achievements Of Ancient Roman Engineers

Water Appia - the main Roman reservoir conduit, built in 312 BC

In 312 BC, a Roman lawmaker Appius Claudius Caecus requested that a reservoir conduit be worked to flexibly Rome with water. The water system was one of two significant Roman activities completed during this period; the other was a street for military purposes.

Among all the accomplishments of antiquated building, Roman reservoir conduits have a place with the most excellent ones.

Specialists in old Roman Empire realized that one of the most key necessities for any town or city to exist is to flexibly it with water, since it is fundamental human need.

The city must have water to drink, cook and clean, however it additionally should have every single irreplaceable course of action to evacuate undesirable water. Both antiquated and present day Rome has been all around provided with the methods for conveying water to the city and removing it.

Antiquated Romans were exceptionally gifted specialists. By the mid fourth century AD, Rome was being provided by in excess of twelve reservoir conduits, which agreeably could bring in excess of a million cubic meters of new water to the city consistently! This gigantic measure of water was conveyed to Rome's occupants through an unpredictable system of tanks and pipes and to around 1,500 open wellsprings, pools and very nearly 900 open and private showers. Understand more photo-1451187580459-43490279c0fa.jpg

7. Incredible Viking Crystal Sunstones Helped Vikings To Navigate

In Viking times the mysterious Sun Stone demonstrated mariners street when the sun went down.

In Viking times the supernatural Sun Stone indicated mariners street when the sun went down.

The Norse adventures notice a secretive "sunstone" - a mystical stone which indicated mariners street when the sun vanished. Presently scientists state the stone is genuine and it's an extraordinary precious stone. One reason why the presences of sunstones have for quite some time been contested is on the grounds that they are contained in the adventure of Saint Olaf, a story with numerous mysterious components. Be that as it may, this has changed and now.

Sunstoncs can never again be viewed as only a legend. Archeologists have found a unique precious stone that proposes incredible Viking sunstones did exists in all actuality. Understand more

8. Antiquated Sophisticated Mercury-Based Gilding That We Still Can't Reach

Antiquated Sophisticated Mercury-Based Gilding That We Still Can't Reach

At times, the innovation was utilized to apply genuine gold and silver. It additionally was utilized falsely, to make modest metal sculptures that resemble strong gold or silver.

People of yore were in control of exceptionally refined information. Old gold and silversmiths utilized mercury, which was created over 8,000 years prior in Turkey. Mercury was utilized for overlaying (arches, insides of church buildings, strict figures and that's only the tip of the iceberg) in numerous pieces of the antiquated world. A large number of the people of yore's methods are as yet obscure. They were gifted to the point that a portion of the quality they accomplished has still not been coordinated. In old occasions, these advanced strategies were utilized to create and enrich various kinds of ancient rarities, for example, gems, sculptures, talismans, and usually utilized articles.

Alchemy Symbols Explained

Gilders played out these procedures not exclusively to improve protests yet in addition to recreate the presence of gold or silver, some of the time deceitfully. From a mechanical perspective, the point of these laborers more than 2000 years prior was to make the valuable metal coatings as slim and follower as could be expected under the circumstances.

This was so as to spare costly metals and to improve the protection from the wear brought about by proceeded with use and course. Ongoing discoveries affirm the significant level of skill came to by the antiquated specialists and experts and stresses an aesthetic nature of the items they created couldn't be bettered in old occasions and has not yet been reached in current ones. Understand more

Chapter-I

Innovation of the Ancients: Artificial Platforms

Innovation of the Ancients in different Artificial Platforms some are

Nan Madol

Two truly momentous islands - Kosrae and Pohnpei ("upon a stone special stepped area") - can be found in the Caroline Islands, a broadly dispersed archipelago situated in the western Pacific Ocean, toward the north of New Guinea and in excess of 1,500 miles toward the east of the Philippines.

These two spots are loaded with puzzles, especially in view of peculiar survives from old design, in type of old massive urban communities, of which beginning nobody can figure."

Those, who don't refute the fantasies and legends of the people of yore, can consider over some intriguing legends and fantasies. One of them says that "enchantment powers" were utilized to move the logs from a distant area, by making them fly but another legend underpins the last mentioned, saying about a relentless "entertainer" who made the logs fly towards Nan Madol, where the city was built and a firebreathing mythical serpent turned into its image.

Who was keen on building two of the Pacific's biggest basalt urban areas? Understand more photo-1485827404703-89b55fcc595e.jpg

Profoundly Advanced Robots In Ancient China

There are numerous instances of robots that were made in antiquated China.

There are numerous instances of robots that were made in old China.

In old China we run over various profoundly propelled robots that could sing, move, act like workers and perform numerous other amazing assignments. A portion of these striking robots are even said to have had life-like organs, for example, bones, muscles, joints, skin and hair. It's a very exceptional considering it is as of late our cutting edge progress has begun to create human-like robots. There is no uncertainty mechanical building in old China arrived at a significant level.

Perusers of Ancient pages know about the mind boggling antiquated machines imagined by Hero of Alexandria, or the exceptional Talos robot, however there are a lot progressively extraordinary instances of old mechanical autonomy found in different corners of the world. History of robots in old China can be followed far back in time. Robots existed not just du

NIMRUD LENS

Nimrud focal point is a 3000-year old bit of rock precious stone that was found in the castle of Nimrud, Iraq by Sir John Layard in 1850. It is supposed to be made during 750-710 BC. Researchers have proposed a hypothesis that this focal point was utilized by antiquated Assyrians as a piece of a telescope, thinking about their insight in space science.

The focal point is somewhat oval and generally ground on a lapidary wheel. It is accepted that the antiquated Assyrians thought about Saturn as a divine being, encompassed by a ring of snakes. This thus demonstrates Galileo wasn't the principal individual to make a telescope.

ANTIKYTHERA MECHANISM

The Antikythera system is an old Greek gadget used to follow the development of the sun, the moon, and the planets to foresee divine occasions and for other mysterious purposes. It was found during the 1990s by jumpers around the shore of the island of Antikythera. Researchers accept that this gadget has a place with the first or second century BC.

The gadget's motivation isn't completely seen, yet its development has perplexed researchers throughout the years. It is regularly alluded to as a simple PC as a result of its capacity to figure sun oriented years and lunar stages.

TELHARMONIUM

The world's first electronic instrument, the Telharmonium was genuinely stand-out. Created by Thaddeus Cahill in 1897, it was a huge melodic organ which delivered imaginative engineered melodic notes by utilizing tonewheels. The sound made was consequently moved to the amplifiers through wires. The instrument is supposed to weigh 200 pounds and sufficiently large to fit a whole room.

After it's underlying achievement, Cahill had large designs for telharmonium. Tragically, the gadget was excessively current for now is the ideal time, devoured a humongous measure of vitality which the early force lattices couldn't deal with, and cost an incredible $200,000.00.

NEPENTHE

An energizer tranquilize, nepenthe was supposed to be a medication of absent mindedness in antiquated Greek writing. It is much of the time referenced in Homer's Odyssey too. While it began in Egypt, the Greek would for the most part treat the dispossessed with this medication. Some contend that the medication is absolutely a craft of fiction, yet some vibe in any case. As a result of its upper properties, it is contrasted with opium.

This is one innovation which is presumably still around today, however analysts can't pinpoint it. A few analysts feel it can either be opium, wormwood concentrate, or scopolamine.

photo-1478432780021-b8d273730d8c.jpg

VITRUM FLEXILE (FLEXIBLE ROMAN GLASS)

Vitrum Flexile was made during the rule of Roman Empiror Tiberius Caesar. It was imagined by Petronius in 63 AD who introduced the vessel to Caesar and requested that he hand it back to him. At the point when Petronius tossed it on the floor, it didn't break however just scratched. The innovator at that point formed it back to its unique shape. Not perceiving the virtuoso, Caesar decapitated the creator dreading depreciation of valuable metals.

Another adaptation of the story told by Dio Cassius depicted the creator as an entertainer. At the point when the vessel was tossed on the floor, it broke, and the creator fixed it with his exposed hands. An organization named Corning acquainted an adaptable glass comparable with Petronius' in 2012.

STRADIVARI VIOLINS

One of the most popular lost advances is the way toward making Stradivari violins. The violins alongside other instruments like violas, cellos, and guitars were made by the Stradivari group of Italy between 1650-1750. The Stradivari violins are a valued belonging today and cost over a hundred thousand dollars. Just 600 of them despite everything stay starting today.

The method to manufacture the Stradivari violins was a privileged bit of information and just known by Antonio Stradivari and his children, Omobono and Francesco. The mystery passed on with them. In the wake of contemplating the violin, scientists have set up a speculation guaranteeing that it is the thickness of a specific wood that delivers the one of a kind sound and reverberation however nobody knows without a doubt.

DAMASCUS STEEL

Utilized in the Middle East around 1100-1700 AD, the Damascus steel was a solid metal. It was for the most part used to make blades and swords. It is known to cut rocks and different metals particularly blades of gentler metals neatly into a half. It is accepted to be made of wootz steel found in Sri Lanka and India.

Blades from Damascus steel were made by joining delicate iron and cementite. The method was lost around 1750 AD. It was most likely in light of the fact that the smiths who made the blades didn't have a particular formula to make them and just went with their impulses.

LYCURGUS CUP

The Lycurgus cup is a glass cup made with dichroic glass that would change hues relying upon the light going through it. This interesting craftsmanship piece uncovers the information old individuals had about nanotechnology. The cup depicts the destruction of Emperor Licinius by Constantine in 308-24 AD.

At the point when light goes through the cup, it changes shading from murky green to a blazing translucent red. The cup contains colloidal gold and silver, giving it these strange optical properties. An exploration group from University of Illinois is attempting to assemble an increasingly refined structure dependent on the science applied to make the Lycurgus cup.

THE VIMANAS

In the Mahabharata messages there are depictions of fight planes that fire rockets that utilization sound to discover their objective and light emissions that devastate anything they contact with their vitality. Credit for these machines was ascribed to the Yavanas who are accepted to be the old Greek civilisations. In or around the mid 1950's a progressively current book was made accessible. Called the Vaimanika Sastra (study of Aeronautics) it was supposedly the "motivated' work of Subbaraya Shastry who guaranteed it depended on the compositions of the extraordinary sage Bharadwaja in this way offering genuineness to the logical cases. The image is an idea attracting dependent on the composed portrayals the writings. In any case, it could similarly well portray the Russian idea rocket delineated on the contiguous stamp. It is likewise important that there are claims that the 1960's Russian researchers took a profound enthusiasm for the Vimana marvel and abnormally it is around this time they made critical jumps forward in their innovative accomplishments. All things considered, this is likely only an incident. Vimana are not remarkable to India and there are references from everywhere throughout the world and incorporate the Egyptian Saqqara Bird, the pre-Columbian brilliant plane models, the Greek Icarus legend, the Chariot of Ezekiel, the Nazca runways (lines), The Abydos carvings, The Tassili rock canvases from Algeria and the Chinese references to Lu Ban's wooden airplane that flew significant stretches. Normally, these references are regularly excused by current history specialists as just incomprehensible yet there can be no uncertainty that humankind has an aggregate memory of have once had the option to fly in old occasions. Is this an incident dependent on the overall unrealistic reasoning of past civilisations or is it a memory of when it was really conceivable. You should choose for yourself.

Most prominent Innovation of All Time

Fire:

In spite of the fact that fire is a characteristic marvel, its disclosure denoted an upheaval in the pages of history. All gratitude to our precursors for driving us to the controlled utilization of fire which encourages us from beautiful lighting to delectable cooking. Likewise, the history of different scenes was adjusted by fire. Antiquated individuals may have gotten familiar with fire got from characteristic sources. Later came the procedures of making fire misleadingly. This wonderful control of fire occurred during Early Stone Age by Homo erectus.

The most punctual proof originates from Kenya district. Despite the fact that fire could have been utilized around 1 million years prior, proof of prepared food is found from 1.9 million years. From the past to the current Fire has been in ceremonies, horticulture, cooking, creating warmth and light, flagging, different modern procedures, incineration, and as a weapon or mode of devastation.

Advancement of Wheel :

The wheel stands apart as the OG of building wonders and one of the most popular creations that affected various different things. This crude innovation made it simpler for us all to travel. From the archeological unearthings, the most seasoned realized wheel is from Mesopotamia, around 3500 B.C. Because of progression in the new and creative plan of wheels, industrialization could flourish. The wheel fills an essential need in our lives, and we were unable to envision the world without them.

Compass:

Made for profound and navigational purposes, the soonest compasses were undoubtedly imagined by the Chinese in around 1050 BC. It was made of lodestones, which is a normally charged iron mineral.

The creation of the electromagnet in 1825 lead to the improvement of the cutting edge compass. The innovation of the compass positively helped current route more than our GPS-requiring society could comprehend.

Car:

Despite the fact that the establishment to the cutting edge vehicle year was laid in 1886 by German creator Karl Benz, Cars didn't turn out to be generally accessible until the mid twentieth century. Henry Ford developed large scale manufacturing procedures that got norm, with Ford, General Motors, and Chrysler. Be that as it may, he positively wasn't the main individual to build up the horseless carriage. The historical backdrop of the car mirrors an overall advancement. Many side project businesses bloomed making a large number of new openings. Oil and steel became two entrenched enterprises. Vehicle creation and deals are one of the significant pointers of the financial status. Additionally, it impacted the innovative advances in oil refining, steel making, paint and fortified glass fabricating, and other modern procedures.

Steam Engine:

Thomas Savery protected the main down to earth steam motor in 1698. It was perhaps the best development made by a man making him one of the individuals who have changed the world. Later in 1781, James Watt protected an improved steam motor and proceeded to fuel one of the most earth shattering mechanical jumps in mankind's history during the Industrial Revolution. During the 1800s these motors lead to an improvement in transportation, agribusiness, and assembling enterprises. Afterward, the steam motor's essential standard set up for developments like interior ignition motors and fly turbines, which provoked the ascent of vehicles and airplane during the twentieth century.

Inside Combustion Engine :

The nineteenth century innovation (made by Belgian architect Etienne Lenoir in 1859 and improved by Germany's Nikolaus Otto in 1876), this motor believers concoction vitality into mechanical vitality overwhelmed the steam motor and is utilized in present day vehicles and planes. Elon Musk's electric vehicle organization Tesla, among others, is as of now attempting to upset innovation in this field indeed

Concrete:

Concrete is the one of the most broadly utilized man-made material. It's a composite material made out of harsh composite reinforced along with a liquid concrete which solidifies after some time. Most cements utilized are lime-based, black-top cement, and polymer cements. Prior, Limestone was utilized as a rough concrete. As the materials and blends improved, current cement was imagined. One of the key elements of cement will be concrete. The establishment to solidify was laid in 1300 BC. Center eastern developers covered the outside of their mud posts with a flimsy, and clammy consumed limestone, which synthetically responded with gasses noticeable all around to shape a hard, defensive surface. Around 6500 BC, the principal solid like structures were worked by the Nabataea merchants or Bedouins in the southern Syria and northern Jordan districts.

By 700 BC, the essentialness of pressure driven lime was known, which prompted the improvement of mortar gracefully ovens for the development of rubble-divider houses, solid floors, and underground waterproof reservoirs. By 3000 BC, the Egyptians were utilizing early types of cement to assemble pyramids. In 1824, the most utilized Portland concrete was concocted by Joseph Aspdin of England. George Bartholomew had set out the primary solid road in the US during 1891, which despite everything exists. Before nineteenth century's over, the utilization of steel-fortified cement was created. In 1902, utilizing steel-strengthened solid, August Perret structured and assembled a high rise in Paris. This structure a wide esteem and notoriety to concrete and furthermore impacted the advancement of fortified cement. In 1921, Eugène Freyssinet spearheaded the utilization of strengthened solid development by building two huge illustrative curved carrier sheds at Orly Airport in Paris.

Petroleum:

Without gas, there couldn't be the primary modern transformation in the car business. Gas is a fuel subordinate of oil, which is in no time called "gas" in the United States and "petroleum," in different spots the world over. To be progressively explicit, petroleum is a straightforward, oil determined fluid that is utilized as a basic fuel in inward burning motors. Petroleum is the characteristic result and the creation here we are discussing is the various procedures to improve the quality. Do you know, gas was at first disposed of? During 1859, in Pennsylvania, Edwin Drake burrowed the main oil well and refined the oil to create lamp oil. In spite of the fact that the refining delivered gas, he disposed of it as he was ignorant of it. Until 1892, the unmistakable quality of gas wasn't perceived. The main gas siphon was made by Sylvanus Bowser On September 5, 1885. The year 1970 picked up consideration towards ecological assurance.

PENICILLIN:

Found by the Scottish researcher Alexander Fleming in 1928, this medication changed medication by its capacity to fix irresistible bacterial sicknesses. It started the period of anti-toxins.

Railroads:

Railroads is a method of transport which can convey countless travelers effortlessly of solace as well as overwhelming burdens to significant distances. Present day trains history is around 200 years of age, which changed the manner in which we travel. Inaccessible grounds become potential, ventures are controlled with an unbounded measure of crude materials. Prior method of transport was trucks pulled by creatures. During 1500 - 1800, wagonways were basic in Europe, which was utilized in mining. After the creation of Steam motor, more specialists were completed all through the world for a superior plan.

The business appearance of train systems came in the late 1820s, and the pioneer in that field was designer George Stephenson, with his structure 'Rocket', the most well known early railroad train. This increased quick extension across recently gained lands. In 1821, Stephenson was selected as an architect for the development of the Stockton and Darlington railroad, which was opened as the primary open railroad in 1825. The stupendous accomplishment of "Rocket" and opening of the Stockton to Darlington railroad line empowered rail line industry. Railroads arrived at another significant section in the history, with the development of Diesel Engine.

Plane:

On December 17, 1903, Wilbur and Orville Wright accomplished the main fueled, supported and controlled plane. While flying machines had been concocted since da Vinci's time, the Wright Brothers turned into the greatest victories. Starting with lightweight flyers, the couple established the framework for present day aeronautical building. Also, new business blasted alongside various individuals being prepared to fly planes. The likelihood to fly more than a large number of miles in less time would not have been made conceivable if the airplane were not imagined.

Nail:

The modern human life would not have been conceivable without the creation of a little nail. They give perhaps the best hint in deciding the period of notable structures. Before the creation of nails, wood structures were worked by geometrically interlocking adjoining sheets. The innovation of nails returns to a few thousand years and was conceivable simply after the advancement of throwing and molding a metal. Around 3400 BC, Bronze nails were found in Egypt. As indicated by the University of Vermont, the hand – fashioned nails were a standard until the 1790s and mid 1800s. By 1913, 90 percent of nails delivered in the U.S. were steel wire nails. Different sorts of nails incorporate pins, tacks, brads, and spikes with wire nails being well known.

OPTICAL LENSES:

from glasses to magnifying instruments and telescopes, optical focal points have incredibly extended the conceivable outcomes of our vision. They have a long history, first created by old Egyptians and Mesopotamians, with key hypotheses of light and vision contributed by Ancient Greeks. Optical focal points were likewise instrumental parts in the production of media advancements associated with photography, film and TV.

Devices:

The utilization of instruments began 2.6 million years back in Ethiopia. Anthropologists accept the utilization of devices turned into a significant advance in the development of humankind. Prior materials, for example, sticks and stones made instruments.

The development of machine instruments propelled the mechanical upheaval. Envision how might we manufacture or keep up past developments without a helpful mallet.

Light:

The vitality we use today at home and office is a splendid thought from over 150 years prior. Spearheaded in the mid nineteenth century by Humphry Davy, electric lights created all through the 1800s and was one of the most powerful, incredible developments all things considered. Edison and Swan protected the main light in 1879 and 1880. In the mid-1980s, CFLs hit the market. Be that as it may, the downsides, for example, significant expense, massive, low light yield, and conflicting execution made them less noticeable. At present, LEDs offer the best vitality investment funds available.

In any case, the innovation of the bulb jolted new organizations. It likewise prompted new vitality discoveries, for example, power plants, electric transmission lines, home machines and so forth.

PAPER :

Designed around 100 BC in China, paper has been fundamental in permitting us to record and offer our thoughts.

Power:

Power has become the fundamental requirement for everyday life. It's been there around from the beginning however the down to earth applications to viably utilize it were concocted. Albeit many use power, how any of you the know the development of power? In , Alessandro Volta found the main viable strategy for creating power. 1831 is denoted the time of significant forward leap for power. A British researcher Michael Faraday found the fundamental standards of power age. The electromagnetic enlistment revelation changed the vitality utilization. Road lights were probably the most punctual consideration picking up hardware. With the ascent in power ease of use, presently it remains as a spine of current mechanical society. With expanded versatility, human life has gotten subject to power.

Battery:

The ancient battery goes back the Parthian domain, which may be 2,000 years of age. The old battery comprised of a dirt container loaded up with a vinegar arrangement, into which a copper chamber encompassed iron pole was embedded. These batteries may have been utilized to electroplate silver.

The designer of the main electric battery is Alessandro Volta. He likewise established the framework of Electrochemistry. The large scale manufacturing of the principal electric battery started in 1802 by William Cruickshank. The historical backdrop of batteries denoted an amazing date in 1859, with the innovation the primary battery-powered battery dependent on lead corrosive by the French doctor Gaston Planté. The Nickel Cadmium (NiCd) battery was presented in 1899 by Waldemar Jungner.

Print machine:

Prior to the Internet's capacity to spread data, the print machine helped data travel all through the globe. Created around 1440 in Mainz, Germany, Johannes Gutenberg's machine enhanced previously existing presses. By 1500 Gutenberg presses were working all through Western Europe with a creation of 20 million duplicates. By 1600, they had made more than 200 million new books.

Morse Code and Telegraph:

The message was created around 1830 – 1840 by Samuel Morse and different creators, which altered significant distance correspondence. The electrical signs were transmitted by a wire laid between stations. What's more, Samuel Morse built up a code, called Morse code, for the straightforward transmission of messages across transmit lines. In view of the recurrence of use, the code alloted a lot of spots (short checks) and runs (long stamps) to English letters in order and numbers. The message established significant frameworks for present day comforts like phones and (a few researchers contend) coding for the Internet.

Steel:

While the early ages utilized stone, bronze, and iron, it was steel that terminated the mechanical upset. According to archeological unearthings, most punctual known creation of the metal goes back to 4,000 years. The development of Bessemer Process (a method for making steel utilizing liquid pig iron) made ready for the large scale manufacturing of steel, making it perhaps the greatest business on earth. Presently steel is utilized in the making of everything from extensions to high rises.

Transistors:

The transistor is a fundamental segment in about each cutting edge electronic contraption. In 1926, Julius Lilienfeld licensed a field-impact transistor, yet the working gadget was not possible. In 1947 John Bardeen, Walter Brattain, and William Shockley built up the main down to earth gadget at Bell Laboratories. It at that point won the trio 1956 Nobel Prize in material science. Transistors have become a key bit of the hardware in incalculable electronic gadgets including TVs, phones, and PCs having a noteworthy effect on innovation.

The Bow and Arrow

The stock long range weapon of armed forces the world over before firearms, bowmen and other rocket troops commanded cautious fighting for centuries.

Anti-toxins:

Anti-toxins spared a huge number of lives by murdering and restraining the development of unsafe microorganisms. Louis Pasteur and Robert Koch originally portrayed the Antibiosis (wonders of anti-infection sedate) in 1877. In 1928, Alexander Fleming set the first jump in quite a while by recognizing penicillin, the substance compound with anti-toxin properties. All through the twentieth century, anti-microbials spread quickly and end up being a significant living improvement, battling almost every known type of contamination and securing people groups' wellbeing.

Contraceptives:

Avoidance of pregnancy has a long and decided history. The historical backdrop of contraceptives goes back to 1500 B.C, where old Egypt ladies would blend nectar, sodium carbonate and crocodile manure into a thick, strong glue called pessary and addition it into their vaginas before an intercourse. Be that as it may, numerous scientists accept that few old world conception prevention techniques are not compelling and in reality could be deadly. The main known type of condom (a goat bladder) was utilized in Egypt around 3000 B.C.

In 1844 Charles Goodyear licensed the vulcanization of elastic, which prompted the large scale manufacturing of elastic condoms. In 1914 with a month to month bulletin called "The Woman Rebel", Margaret Sanger, extraordinary female instructor from New York state, first authored the "Conception prevention" express. Afterward, Carl Djerassi had effectively made a progesterone pill, which could square ovulation. The Pill raised a worldwide insurgency and was a colossal hit.

X-beam:

Obviously, x-beams are a marvel of the common world, and therefore can't be designed. In any case, they were found inadvertently. The imperceptible was made obvious in 1895. X-beam is without a doubt one of the age making progression in the field of medication. All credits to physicist Wilhelm Conrad Rontgen. While testing whether cathode beams could go through glass, he saw a sparkle originating from a close by artificially covered screen. In view of their obscure nature, he named it as X-beams. Through his perception, he discovered that X-beams can be captured when they enter into human tissue. In 1897, during the Balkan war, X-beams were first used to discover projectiles and broken bones inside patients. In 1901, he got Nobel prize in material science for his work.

Fridge:

In the course of the most recent 150 years, refrigeration offered us approaches to protect food, prescriptions, and other transient substances. Prior to its origination, individuals cooled their food with ice and day off. James Harrison manufactured the principal commonsense fume pressure refrigeration framework. In any case, the primary far reaching cooler was the General Electric "Screen Top" fridge of 1927. While it assisted with firing up mechanical procedures, it turned into an industry itself.

TV:

TV! A little box with tremendous data that changed diversion and interchanges for eternity. The innovation of the TV was crafted by numerous people. In spite of the fact that TV has a significant influence of our regular day to day existences, it quickly created during the nineteenth and the twentieth century. The primary TV camera was developed by two men without realizing that the two of them are chipping away at a similar one (No TV to impart them the news); Vladimir Zworykin and Philo Taylor. In 1884, Paul Gottlieb Nipkow made and protected the primary TV which he called the electromechanical TV framework.

Despite the fact that Color TV was not another thought, in 1925 Zworykin documented a patent for an electronic shading TV framework. After some time, TV will increase political significance as each nation began to share their political plan through it. Television likewise turned into a food method of saving harmony and request.

Camera:

The camera is without a doubt one of the most esteemed manifestations. Cameras have seen numerous periods of development – camera obscura, daguerreotypes, dry plates, calotypes, film to SLRs and DSLRs. In 1826, Joseph Nicéphore Niépce utilized a sliding wooden box camera made by Charles and Vincent Chevalier to tap the primary perpetual photo. With the innovative headways, Digital cameras were acquainted with spare pictures on the memory cards as opposed to utilizing films.

The historical backdrop of the advanced camera started with Eugene F. Lally thought to take photos of the planets and stars while going through space. Afterward, Steven Sasson a Kodak engineer developed and manufactured the primary computerized camera in 1975. In spite of the fact that the computerized camera managed over the customary camera, the most progressive perspective has been the appearance of the camera telephone. Presently, every cell phone has an inbuilt camera and can take pictures. With the developing interest, video recording was likewise made a piece of it. At present, the camera accompanies inbuilt GPS framework and constant geotagging alternatives. Freeze the incredible minutes from your life as photos with better quality and unrivaled taking care of computerized camera. One doesn't need to look a lot farther than a photograph collection to see that cameras are one of the extraordinary innovations that changed the world.

The Radio :

The innovation of the radio was the first among all the significant developments of this time of 100 years (1900-2000). Radio was not developed by a solitary individual, yet was come about out of commitments of a few researchers and designers. Despite the fact that professed to be concocted before, patent for radio was conceded in 1904. Both, Nikola Tesla and Guglielmo Marconi are considered as innovators of the Radio. In 1901, Marconi sent the main radiotelegraph message from England to Canada. Radio-telecommunication got one of the most utilized correspondence methods after the radio was imagined. The AM (Amplitude Modulation) radio was a change that caused in making access to various radio broadcasts conceivable.

Wellspring Pen :

The Romanian creator Petrache Poenaru got a French patent on May 25, 1827, for the development of the main wellspring pen with a barrel produced using an enormous swan plume.

PC:

Major holler to the mechanical architect Charles Babbage for establishing the framework to this amazing and most dependable creation. In the mid nineteenth century, the "father of the PC" conceptualized and designed the principal mechanical PC. In spite of the fact that there's no single creator of the advanced PC, the rule was proposed by Alan Turing in his fundamental 1936 paper. Today, PCs remain as the emblematic portrayal of the cutting edge world.

Email:

During 1969, soon after the making of ARPANET, test email moves between discrete PC frameworks started. Prior to email, sending a significant report abroad includes a turbulent procedure. Presently correspondence is a single tick away. The principal generous utilization of email started during the 1960s. By mid-1970s, it had taken the perceived structure. The present-day the greater part of the official business correspondence relies upon email. Presently, email is accessible on bounty well disposed web interfaces by suppliers, for example, Gmail, Outlook, Yahoo, Hotmail, and so forth. This great vehicle of correspondence is all around received by a huge number of individuals.

Web:

In contrast to the bulb or the phone, the Internet has no single "creator." Instead, it has developed after some time. It began in the United States around the 1950s, alongside the advancement of PCs. Since the mid-1990s, the Internet has revolutionarily affected innovation, including the ascent of electronic mail, texting, voice over Internet Protocol (VoIP) calls, and two-way intelligent video calls.

Internet:

The Internet is a systems administration framework. Though the World Wide Web is an approach to get to data over the mode of the Internet. The dad of the World Wide Web is a British Computer Scientist, Tim Berners-Lee. While filling in as a product engineer at CERN in Geneva, Switzerland, Tim saw that the trouble in sharing data. In 1989, this prompts a proposition "Data Management: A Proposal". In any case, it was not promptly acknowledged. By October 1990, three Tim established the framework to the web through HTML, URL, and HTTP innovations. April 1993, denoted a significant advance throughout the entire existence of Web. The choice to utilize the web for nothing was reported. Right up 'til today, the Web shined an all encompassing influx of innovativeness. The web quickly changed the conventional way and affected the advancement of different enterprises. For instance, it prompted the advancement of online training and economy; the most ideal approach to advance your organization in 2017 is through Google search. Individuals can peruse or observe any kinds of substance online whether through a website or internet based life, for example, Facebook and Twitter.

Banknote:

From materials like animals to valuable metals and coins, money took different structures since the commencement. Because of successive deficiencies of coins, banks gave paper notes as a guarantee against installment of valuable metals in future. Using a light-weight substance as cash began in China during the Han Dynasty in 118 BC. Through explorers, Europe was acquainted with this framework in the thirteenth century. The change to paper cash assuaged governments during emergency time. Along these lines, it changed the substance of the worldwide economy with an imperative advance in another money related framework.

Charge cards:

During the beginning of twentieth century, individuals paid for everything with money. Credit cards began around 1950 by Ralph Schneider and Frank McNamara, organizers of Diners Club, to combine different cards. While innovation keeps on propelling, paying for every day buys with a card has now become the standard.

ATM:

The development of ATM (Automated Teller Machine) is among the most significant innovations at any point made. In the current world, ATMs directed the banking into another idea of self-administration. As indicated by the ATM Industry Association (ATMIA), there are presently over 2.2 million ATM machines introduced around the world. Utilizing ATM, clients make an assortment of exchanges, for example, money withdrawals, check adjusts, or credit cell phones. Numerous specialists accept that the main ATM was the production of Luther Simjian, called Bankograph. In 1967, John Shepherd-Barron concocted a brilliant thought of cash candy machine, which was executed a London bank called Barclays. Prior machines utilized paper vouchers rather than plastic cards. The client entered a distinguishing proof code and can draw a limit of £10 at once. Dallas Engineer Donald Wetzel concocted the principal mechanized financial machine in the U.S.

photo-1455165814004-1126a7199f9b.jpg

Phone and Mobile Phones:

"Mr. Watson, come here, I need you." On March 10, 1876, these were the primary words verbally expressed by phone designer Alexander Graham Bell through his gadget to his associate Thomas Watson. Phone history possibly began with the human want to convey far and wide. With the appearance of the cell phone during the 1980s, correspondences were not, at this point limited. The astute development of cell organize bolstered the transformation of the phone business. Beginning from massive mobiles telephones to ultrathin handsets, mobiles telephones have secured far up until this point. John F. Mitchell and Martin Cooper of Motorola exhibited the main handheld gadget in 1973. Researchers keep on making new thoughts that will additionally support clients.

Robot:

Mechanical gadgets frequently perform muddled, dull, and here and there hazardous undertakings. The word Robot brings out different gadgets extending from a cooking gadget to the Rover. "Robot" first showed up in R.U.R. (Rossum's Universal Robots), a play was composed by Czech writer Karl Capek in 1921. Unintentionally, "apply autonomy" was likewise begat by a sci-fi essayist Isaac Asimov in his short story "Runabout", distributed in 1942. Around 3000 B.C, human dolls were utilized to strike the hour chimes in the Egyptian water tickers. This denoted the principal mechanical plan. As the time flew, more structures and gadgets were advanced. Yet, Robotics progressed logically in the twentieth century.

The establishment to present day robots was laid during the 1950s by George C. Devol, who designed and protected a reprogrammable controller called "Unimate," from "General Automation." In the late 1960s, Joseph Engleberger obtained the patent and changed them into Industrial robots. This exertion made him "the Father of Robotics." Who knows! Some time or another robots may outfox us and make us totally innovation wards. They are genuinely developments that changed the world!

Weapons:

For certain weapons may be a shocking creation while for other people, it may be a horrible innovation. Weapons have been the essential devices since antiquated age. In any case, the Guns have reformed the world. The most punctual use of a gun may have been in China during the thirteenth century CE. In prior days, firearms were terminated by holding a consuming wick to a "contact gap" in the barrel touching off the powder inside. The principal mechanical firearm is the matchlock, which dates to 1400s. By the twelfth century, the innovation began spreading to Asia, trailed by Europe. The issue of stacking and unwavering quality was understood by the development of a hand-driven automatic rifle called Gatling weapon. It was concocted by Richard J. Gatling during the American Civil War. As the tech kept on developing, each after model turned out to be all the more dangerous.

DNA :

The disclosure of DNA totally changed our origination of how we work. As we map the genome and make new disclosures day by day, our comprehension of hereditary qualities will totally change our idea of living – and of death.

Movies:

Nearly everybody wants to watch films of different sorts like a romantic tale, satire, dramatization, repulsiveness, anticipation, activity, fiction, life story and so on. A film is additionally called a film, movie, showy film, photoplay, flick. The name "film" starts from the way that a photographic film has been the vehicle for recording and showing movies. An Early motivation for motion pictures were the plays and move, which had components normal to movie: contents, sets, outfits, creation, course, on-screen characters, crowds, and storyboards. Later in the seventeenth century, the lamps were utilized to extend activity, which was accomplished by different kinds of mechanical slides.

A lot later in 1839, Henry Fox Talbot makes a significant progression in photography creation. The year 1846 was significant for the advancement of movies. The primary film at any point made is the pony moving. In March 1895, the primary film with a Cinématographe camera was shot on La Sortie de leucine Lumière a Lyon (Workers leaving the Lumière processing plant at Lyon). With time, the motion pictures developed with sound, music, shading, and cutting edge innovation.

photo-1535223289827-42f1e9919769.jpg

AI :

The historical backdrop of the field of Machine Learning is an intriguing story. In 1946 the principal PC framework ENIAC was created. Around then the word 'PC' implied an individual that performed numerical calculations on paper and ENIAC was known as a numerical registering machine. This machine was physically worked, for example a human would make associations between parts of the machine to perform calculations. The thought around then was that human reasoning and learning could be rendered legitimately in such a machine.

In 1950 Alan Turing proposed a test to gauge its presentation. The Turing test depends on the possibility that we can possibly decide whether a machine can really learn in the event that we speak with it and can't recognize it from another human. In spite of the fact that, there have not been any frameworks that finished the Turing assessment many fascinating frameworks have been created.

Development and Technology: The Irreplaceable Human Factor

Development is something beyond a popular expression today — it's beginning and end. On the off chance that associations aren't viewed as creative, they're considered additionally rans.

The issue with this attitude? Advancement is regularly compared with the most current types of innovation, as though development can't occur without an immense help from tech.

Here's our world: Innovation can — and must — happen without being driven by innovation. People must keep on concocting huge, new thoughts while in the shower, on the backs of napkins, and, above all, with one another in one-on-one, human-to-human discussions.

It's this low-tech advancement that makes a portion of our best thoughts. Truly, innovation quite often turns into the foundation of these new ideas. Be that as it may, the development didn't begin there. It began with a revelation on the commute home, in a gathering when an "aha" second precipitously happened and in snapshots of calm reflection.

Since at last, innovation is about the "what" and the "how" — and not the topic of "why."

Concentrate on the Why

The "why" is the thing that drives us.

Our "why" gives us reason, inspiration and even expectation. Our "why" causes us remain concentrated on what is important. It empowers us to beat hindrances to progress. It motivates us to strive, to turn when essential and to wrap up.

It appears as though the more innovation we have accessible, the more corners we need to cut. Building something quicker, taller and greater appears to be a higher priority than beginning with a strong "why" establishment. Therefore, we appear to be less ready to buckle down. Awfully frequently en route, we overlook our "why."

Innovation can drastically improve what we do and how we do it. However, it can't contact our "why" — which is a simply human driver of activity.

Is Technology Always a Good Business Partner?

Without a doubt, we can utilize man-made consciousness to assist us with settling on proof based choices — and we

Like Slack and Zoom. Programming as a Service and the cloud make it simpler than any time in recent memory to team up with individuals everywhere throughout the world.

Yet, in the event that we don't keep up the right "why," the innovation doesn't help. See, for instance, at what candidate following frameworks accomplished for the competitor experience. Certainly, they made the lives of managers and spotters simpler. In any case, that "simpler" frequently came to the detriment of bosses' brands. While positively creative in idea, was ATS incredible innovation? Was it the correct sort of "why"? Or on the other hand, regardless of the best goals, did ATS cross into the mechanical clouded side?

Chapter-II

People and Technology

I am a recuperating Silicon Valley engineer who, in 1984, had a first-of-its-sort IBM PCjr around my work area. A couple of years after the fact, I purchased a Macintosh before a great many people had even known about Apple. I gladly claimed one of the first (and greatest, I concede) mobile phones at any point made by Motorola, and was excited with my pull cup-mounted Magellan GPS. Having worked in a virtual limit since 1999, I have grasped new computerized, systems administration and correspondence development and innovation every step of the way. I'm not actually a Luddite.

So I realize that people will and should use innovation. I comprehend that "people and innovation" is truly an "and" explanation, not an "or" proclamation.

photo-1581091226825-a6a2a5aee158.jpg

But then, I realize innovation isn't generally the appropriate response. Many new businesses and Fortune 500 organizations have taken on representative commitment — yet 30 years and billions of dollars later we haven't improved worker commitment the slightest bit. We've planned the best chatbots accessible, however clients can without much of a stretch leave those mechanized discussions unfulfilled. We've had a persistent spotlight on organization culture for as long as decade, with numerous innovation stages intended to assist us with checking and improve both culture and atmosphere in the work environment — yet most by far of representatives despite everything rate their corporate culture as poor.

Innovation Can't Replace Innate Human Characteristics

In every one of the situations above, it isn't that the innovation was inadequately structured or executed. What's more, it isn't that individuals weren't thankful for the help from innovation. It's that they missed what makes a difference most. Pioneers and directors who care enough to truly connect with representatives — not exactly when provoked to by an application. Individual people who need to assemble human-to-human connections, regardless of how fleeting, sufficiently able to satisfy clients. Also, CXOs and colleagues who set out to assemble and put resources into organization culture where individuals feel esteemed and have a feeling of having a place.

It is these basic human qualities — mindful, relationship-assembling and conscious top-down thoughtfulness regarding what is important most — that make our associations "imaginative."

Since regardless of the amount we depend on innovation to take care of business, it's only in the interest of personal entertainment (except if we're talking self-sufficient vehicles). Innovation serves us well as the conveyance instrument. Yet, it isn't the distinction creator.

Individuals give that flash of advancement. We are the key human factor.

Indeed, even at a beginning phase humanity endeavored to assemble increasingly elevated. We expand on an absurd scale and burn through thousands or a great many long periods of work on a solitary bit of structure, which could possibly be inclined to quakes and different assaults of time.

A portion of our most noteworthy structures are entirely old and hard to decide the exact dates they were fabricated.

We despite everything haven't the foggiest how the Great Pyramid of Giza was manufactured, or its exact reason. Regardless of what you may have heard, no mummy has ever been found in the Egyptian pyramids, they were totally found in the Valley of the Kings. So the genuine importance of the pyramids is really a puzzle.

Similar to the innovation utilized and the matter of how numerous societies in Africa, the Middle East and Central America every fabricated pyramid around generally a similar time. The issue has had archeologists both expert and novice the same scratching their heads and conjecturing why and how this could have occurred.

For what reason do we try to such great statures? Conscience maybe. At times we probably won't have a lot of decision yet to assemble upwards if the populace becomes thick and land near water and food is scant.

We flourish in the absolute most unfriendly places on the Earth, and consistently we fabricate upwards.

There is no exact start for the historical backdrop of design either. Our soonest structures date from either the finish of the last ice age or during the ice age, which was just 10 to 15 milleniums back.

In like manner, there was no exact completion of the ice age. We assume it eliminated gradually, however it could have changed rapidly very quickly or years. We truly don't have the foggiest idea. It was a period of emotional changes, monstrous floods and seismic tremors.

Such emotional quakes that individuals in two separate pieces of the world (Egypt and Bolivia) began building tremor safe structures that despite everything stand today. Gigantic Island in Egypt and the Ruined City of Tiahuanacu in Bolivia utilized indistinguishable procedures to safely affix the stones in their structures and make the general structure progressively impenetrable to time.

Pyramids are the prime case of that spearheading human soul to fabricate something indestructible, and the most punctual pyramids are not Egyptian, yet were rather worked in Mesopotamia and Zimbabwe.

The reality the individuals of Zimbabwe began building pyramids initially is amazingly intriguing. Africa was after all the introduction of human progress. It is there we locate the most seasoned enduring structures and the start of our goal to fabricate higher.

The Greeks talked about Mount Olympus and endeavored to copy the divine beings by expanding on head of mountains. The people groups of the Middle East manufactured monstrous Ziggurat step pyramids and enlivened the tale of the Tower of Babel.

We can just accept that the early individuals who fabricated towers of stone in Zimbabwe had a strict or even logical thinking behind what they were building.

At the point when we discuss such structures we can't disregard the logical perspective. These were clearly societies with an enthusiasm for designing, science and investigating the limits of what they could manufacture.

Whether or not it was a sanctuary, a castle, an open air theater for games, an amphitheater for sensational exhibitions and legislative issues, there was consistently that basic designing and innovative soul.

All they required in truth was the hands to convey the stones, the instruments to cut the stones, the splendor of their specialists or more all else:

Engineering Timeline - Western Influences on Building Design

When did Western engineering start? Some time before the eminent structures of antiquated Greece and Rome, people were planning and developing. The period known as the Classical Era developed from thoughts and development strategies that advanced hundreds of years and ages separated in far off areas.

photo-1584457021185-49623a766992.jpg

This survey represents how each new development expands on the one preceding. Despite the fact that our course of events records dates related generally to American design, notable periods don't begin and stop at exact focuses on a guide or a schedule. Periods and styles stream together, at times combining opposing thoughts, now and then creating new methodologies, and regularly re-arousing and re-imagining more seasoned developments. Dates are consistently surmised—engineering is a liquid workmanship.

11,600 BCE to 3,500 BCE — Prehistoric Times

flying perspective on dissipated gigantic stones dispersed around **Stonehenge in Amesbury, United Kingdom.**

Archeologists "burrow" ancient times. Göbekli Tepe in present day Turkey is a genuine case of archeological engineering. Prior to written history, people built earthen hills, stone circles, stone monuments, and structures that regularly puzzle advanced archeologists. Ancient design incorporates momentous structures, for example, Stonehenge, precipice homes in the Americas, and cover and mud structures lost to time. The beginning of design is found in these structures.

Ancient developers moved earth and stone into geometric structures, making our soonest human-made arrangements. We don't have the foggiest idea why crude individuals started fabricating geometric structures. Archeologists can just conjecture that ancient individuals looked to the sky to emulate the sun and the moon, utilizing that round shape in their manifestations of earth hills and solid henges.

Many fine instances of very much saved ancient engineering are found in southern England. Stonehenge in Amesbury, United Kingdom is a notable case of the ancient stone circle. The close by Silbury Hill, additionally in Wiltshire, is the biggest man-made, ancient earthen hill in Europe. At 30 meters high and 160 meters wide, the rock hill is layers of soil, mud, and grass, with burrowed pits and passages of chalk and clay.1 Completed in the late Neolithic time frame, roughly 2,400 BCE, its modelers were a Neolithic human progress in Britain.

The ancient locales in southern Britain (Stonehenge, Avebury, and related destinations) are on the whole an UNESCO World Heritage Site. "The plan, position, and between relationship of the landmarks and destinations," as per UNESCO, "are proof of a rich and profoundly sorted out ancient society ready to force its ideas on the earth." To a few, the capacity to change the earth is key for a structure to be called design. Ancient structures are now and then thought about the introduction of engineering. In the case of nothing else, crude structures surely bring up the issue, what is engineering?

For what reason does the circle rule man's most punctual engineering? It is the state of the sun and the moon, the main shape people acknowledged to be critical to their lives. The pair of engineering and geometry goes path back in time and might be the wellspring of what people find "wonderful" even today.

3,050 BCE to 900 BCE — Ancient Egypt

Blue sky, huge earthy colored pyramid close to street and little individuals and camel figures

The Pyramid of Khafre (Chephren) in Giza, Egypt.

In old Egypt, incredible rulers developed amazing pyramids, sanctuaries, and altars. A long way from crude, tremendous structures, for example, the Pyramids of Giza were accomplishments of building equipped for arriving at extraordinary statures. Researchers have portrayed the times of history in antiquated Egypt.

Wood was not broadly accessible in the parched Egyptian scene. Houses in old Egypt were made with squares of sun-prepared mud. Flooding of the Nile River and the assaults of time devastated the vast majority of these antiquated homes. Quite a bit of what we think about antiquated Egypt depends on incredible sanctuaries and burial chambers, which were made with rock and limestone and improved with hieroglyphics, carvings, and splendidly shaded frescoes. The antiquated Egyptians didn't utilize mortar, so the stones were painstakingly sliced to fit together.

The pyramid structure was a wonder of designing that permitted antiquated Egyptians to construct huge structures. The advancement of the pyramid structure permitted Egyptians to assemble colossal burial places for their rulers. The inclining dividers could arrive at incredible statures in light of the fact that their weight was bolstered by the wide pyramid base. A creative Egyptian named Imhotep is said to have structured one of the soonest of the huge stone landmarks, the Step Pyramid of Djoscr (2,667 BCE to 2,648 BCE).

Developers in antiquated Egypt didn't utilize load-bearing curves. Rather, sections were set near one another to help the substantial stone entablature above. Brilliantly painted and extravagantly cut, the segments regularly copied palms, papyrus plants, and other plant structures. Throughout the hundreds of years, in any event thirty particular segment styles advanced. As the Roman Empire involved these terrains, both Persian and Egyptian segments have affected Western engineering.

Archeological disclosures in Egypt stirred an enthusiasm for the antiquated sanctuaries and landmarks. Egyptian Revival design got elegant during the 1800s. In the mid 1900s, the revelation of King Tut's burial chamber mixed an interest for Egyptian ancient rarities and the ascent of Art Deco design.

850 BCE to CE 476 — Classical

Old roman structure with segments and pediment patio with huge arch behind

The Pantheon, A.D. 126, Rome, Italy

Old style engineering alludes to the style and plan of structures in antiquated Greece and old Rome. Old style engineering formed our way to deal with working in Western states the world over.

From the ascent of antiquated Greece until the fall of the Roman realm, extraordinary structures were developed by exact guidelines. The Roman engineer Marcus Vitruvius, who lived during first century BCE, accepted that developers should utilize scientific standards while building sanctuaries. "For without evenness and extent no sanctuary can have a standard arrangement," Vitruvius wrote in his well known treatise De Architectura, or Ten Books on Architecture.

In his compositions, Vitruvius presented the Classical requests, which characterized segment styles and entablature plans utilized in Classical design. The soonest Classical requests were Doric, Ionic, and Corinthian.

In spite of the fact that we consolidate this building time and call it "Old style," students of history have depicted these three Classical periods:

700 to 323 BCE — Greek: The Doric section was first evolved in Greece and it was utilized for extraordinary sanctuaries, remembering the celebrated Parthenon for Athens. Basic Ionic segments were utilized for littler sanctuaries and building insides.

323 to 146 BCE — Hellenistic: When Greece was at the tallness of its capacity in Europe and Asia, the domain fabricated expand sanctuaries and common structures with Ionic and Corinthian segments. The Hellenistic time frame finished with successes by the Roman Empire.

44 BCE to 476 CE — Roman: The Romans acquired vigorously from the previous Greek and Hellenistic styles, yet their structures were all the more profoundly ornamented. They utilized Corinthian and composite style sections alongside enriching sections. The creation of cement permitted the Romans to assemble curves, vaults, and arches. Popular instances of Roman engineering remember the Roman Colosseum and the Pantheon for Rome.

Quite a bit of this antiquated design is in ruins or somewhat remade. Computer generated reality programs like Romereborn.org endeavor to carefully reproduce the earth of this significant human progress.

527 to 565 — Byzantine

red stone holy structure with chamber focus arch and numerous rooflines

Church of Hagia Eirene in the First Courtyard of the Topkapı Palace, Istanbul, Turkey.

After Constantine moved the capital of the Roman domain to Byzantium (presently called Istanbul in Turkey) in 330 CE, Roman design developed into a smooth, traditionally motivated style that pre-owned block rather than stone, domed rooftops, expand mosaics, and old style structures. Head Justinian (527 to 565) drove the way.

Eastern and Western conventions consolidated in the hallowed structures of the Byzantine time frame. Structures were planned with a focal arch that inevitably rose higher than ever by utilizing building rehearses refined in the Middle East. This time of building history was transitional and transformational.

800 to 1200 — Romanesque

Adjusted curves, enormous dividers, tower of the Basilica of St. Sernin (1070-1120) in Toulouse, France

Romanesque Architecture of the Basilica of St. Sernin (1070-1120) in Toulouse, France.

As Rome spread across Europe, heavier, stocky Romanesque design with adjusted curves rose. Houses of worship and palaces of the early Medieval period were developed with thick dividers and overwhelming wharfs.

Indeed, even as the Roman Empire blurred, Roman thoughts came to far across Europe. Worked somewhere in the range of 1070 and 1120, the Basilica of St. Sernin in Toulouse, France is a genuine case of this transitional design, with a Byzantine-domed apse and an additional Gothic-like steeple. The floor plan is that of the Latin cross, Gothic-like once more, with a high change and tower at the cross convergence. Built of stone and block, St. Sernin is on the journey course to Santiago de Compostela.

1100 to 1450 — Gothic

Engineering Reaches New Height Built in the thirteenth century, Chartres Cathedral in Chartres, France is a showstopper of Gothic Architecture

The Gothic Cathedral of Notre Dame de Chartres, France.

From the get-go in the twelfth century, better approaches for building implied that houses of God and other enormous structures could take off higher than ever. Gothic engineering became described by the components that upheld taller, increasingly agile design—developments, for example, pointed curves, flying supports, and ribbed vaulting. What's more, expound recolored glass could replace dividers that never again were utilized to help high roofs. Figures of grotesqueness and other chiseling empowered pragmatic and beautifying capacities.

A significant number of the world's most notable hallowed spots are from this period in structural history, including Chartres Cathedral and Paris' Notre Dame Cathedral in France and Dublin's St. Patrick's Cathedral and Adare Friary in Ireland.

Gothic design started chiefly in France where developers started to adjust the previous Romanesque style. Developers were likewise impacted by the sharp curves and expound stonework of Moorish design in Spain. One of the most punctual Gothic structures was the mobile of the nunnery of St. Denis in France, worked somewhere in the range of 1140 and 1144.

Initially, Gothic design was known as the French Style. During the Renaissance, after the French Style had dropped outdated, craftsmans taunted it. They instituted the word Gothic to propose that French Style structures were the rough work of German (Goth) savages. Despite the fact that the name wasn't exact, the name Gothic remained.

While developers were making the incomparable Gothic houses of prayer of Europe, painters and stone workers in northern Italy were splitting endlessly from inflexible medieval styles and establishing the framework for the Renaissance. Workmanship antiquarians consider the period between 1200 to 1400 the Early Renaissance or the Proto-Renaissance of craftsmanship history.

Interest for medieval Gothic engineering was stirred in the nineteenth and twentieth hundreds of years. Planners in Europe and the United States structured incredible structures and private homes that imitated the houses of God of medieval Europe. On the off chance that a structure looks Gothic and has Gothic components and attributes, however it was worked during the 1800s or later, its style is Gothic Revival.

1400 to 1600 — Renaissance

stone estate on a provincial slope, square with four colonnades on each side, focus vault, balanced

Estate Rotonda (Villa Almerico-Capra), close to Venice, Italy, 1566-1590, Andrea Palladio. Massimo Maria Canevarolo by means of Wikimedia Commons, Creative Commons Attribution-ShareAlike 3.0 Unported (CC BY-SA 3.0)

An arrival to Classical thoughts guided a "time of enlivening" in Italy, France, and England. During the Renaissance period planners and developers were roused by the deliberately proportioned structures of antiquated Greece and Rome. Italian Renaissance ace Andrea Palladio stirred an enthusiasm for old style engineering when he planned delightful, profoundly balanced manors, for example, Villa Rotonda close to Venice, Italy.

Chapter-III

Extraordinary Ancient and Modern Architecture

Over 1,500 years after the Roman planner Vitruvius composed his significant book, the Renaissance draftsman Giacomo da Vignola sketched out Vitruvius' thoughts. Distributed in 1563, Vignola's The Five Orders of Architecture turned into a guide for manufacturers all through western Europe. In 1570, Andrea Palladio utilized the new innovation of portable kind to distribute I Quattro Libri dell' Architettura, or The Four Books of Architecture. In this book, Palladio indicated how Classical guidelines could be utilized for great sanctuaries as well as for private manors.

Palladio's thoughts didn't emulate the Classical request of engineering however his structures were in the way of old plans. Crafted by the Renaissance experts spread across Europe, and long after the time finished, draftsmen in the Western world would discover motivation in the flawlessly proportioned design of the period. In the United States its relative plans have been called neoclassical.

1600 to 1830 — Baroque

luxurious access to The Palace of Versailles in France

The Baroque Palace of Versailles in France.

Ahead of schedule during the 1600s, an expand new engineering style showered structures. What got known as Baroque was described by complex shapes, unrestrained adornments, rich artistic creations, and intense differences.

In Italy, the Baroque style is reflected in lavish and sensational places of worship with sporadic shapes and excessive ornamentation. In France, the profoundly ornamented Baroque style joins with Classical limitation. Russian blue-bloods were intrigued by the Palace of Versailles, France and fused Baroque thoughts in the structure of St. Petersburg. Components of the detailed Baroque style are found all through Europe.

Engineering was just a single articulation of the Baroque style. In music, popular names included Bach, Handel, and Vivaldi. In the workmanship world, Caravaggio, Bernini, Rubens, Rembrandt, Vermeer, and Velázquez are recalled. Well known creators and researchers of the day incorporate Blaise Pascal and Isaac Newton.

1650 to 1790 — Rococo

lavish royal residence, flat direction, blue veneer, wide street prompting lined section

Catherine Palace Near Saint Petersburg, Russia. Saravut Eksuwan/Getty Images

During the last period of the Baroque time frame, manufacturers built smooth white structures with clearing bends. Lavish workmanship and engineering is portrayed by exquisite enhancing structures with scrolls, vines, shell-shapes, and sensitive geometric examples.

Lavish modelers applied Baroque thoughts with a lighter, increasingly smooth touch. Actually, a few history specialists recommend that Rococo is basically a later period of the Baroque time frame.

Designers of this period incorporate the incomparable Bavarian plaster aces like Dominikus Zimmermann, whose 1750 Pilgrimage Church of Wies is an UNESCO World Heritage site.

1730 to 1925 — Neoclassicism

Huge even situated arrangement of associated structures with an arch in the middle

The U.S. State house in Washington, D.C.

By the 1700s, European draftsmen were getting some distance from expand Baroque and Rococo styles for controlled Neoclassical methodologies. Organized, even Neoclassical design mirrored the scholarly arousing among the center and high societies in Europe during the period students of history regularly call the Enlightenment. Elaborate Baroque and Rococo styles become undesirable as draftsmen for a developing working class responded to and dismissed the lavishness of the decision class. French and American upheavals returned plan to Classical beliefs—including equity and popular government—significant of the civic establishments of old Greece and Rome. A distinct fascination for thoughts of Renaissance designer Andrea Palladio propelled an arrival of Classical shapes in Europe, Great Britain, and the United States. These structures were proportioned by the old style orders with subtleties obtained from antiquated Greece and Rome.

In the late 1700s and mid 1800s, the recently framed United States attracted upon Classical beliefs to develop terrific government structures and a variety of littler, private homes.

1890 to 1914 — Art Nouveau

corner perspective on huge, multi-story lodging with dormers and overhangs with created iron rails whirls

The 1910 Hôtel Lutetia in Paris, France

Known as the New Style in France, Art Nouveau was first communicated in quite a while and visual communication. The style spread to design and furniture during the 1890s as a rebel against industrialization directed individuals' concentration toward the regular structures and individual craftsmanship of the Arts and Crafts Movement. Craftsmanship Nouveau structures regularly have unbalanced shapes, curves, and embellishing Japanese-like surfaces with bended, plant-like plans and mosaics. The period is regularly mistaken for Art Deco, which has an altogether unique visual look and philosophical inception.

photo-1551563530-a7660dea9972.jpg

Note that the name Art Nouveau is French, yet the way of thinking—somewhat spread by the thoughts of William Morris and the works of John Ruskin—offered ascend to comparative developments all through Europe. In Germany it was called Jugendstil; in Austria it was Sezessionsstil; in Spain it was Modernismo, which predicts or occasion starts the advanced time. Crafted by Spanish modeler Antoni Gaudí

(1852–1926) are supposed to be affected by Art Nouveau or Modernismo, and Gaudi is frequently called one of the main pioneer planners.

1895 to 1925 — Beaux Arts

profoundly luxurious outside of rectangular box-formed structure with curves and sections and figures lit around evening time

The Paris Opera by Beaux Arts

Otherwise called Beaux Arts Classicism, Academic Classicism, or Classical Revival, Beaux Arts engineering is described by request, balance, formal structure, pretentiousness, and expand ornamentation.

Consolidating traditional Greek and Roman engineering with Renaissance thoughts, Beaux Arts design was a supported style for excellent open structures and rich chateaus.

1905 to 1930 — Neo-Gothic

Detail of the head of a lavishly cut out high rise in Chicago

The Neo-Gothic 1924 Tribune Tower in Chicago

In the mid twentieth century, medieval Gothic thoughts were applied to current structures, both private homes and the new sort of engineering called high rises.

Gothic Revival was a Victorian style motivated by Gothic church buildings and other medieval design. Gothic Revival home structure started in the United Kingdom during the 1700s when Sir Horace Walpole chose to redesign his home, Strawberry Hill. In the mid twentieth century, Gothic Revival thoughts were applied to present day high rises, which are regularly called Neo-Gothic. Neo-Gothic high rises frequently have solid vertical lines and a feeling of extraordinary stature; angled and pointed windows with enlivening tracery; foreboding figures and other medieval carvings; and zeniths.

The 1924 Chicago Tribune Tower is a genuine case of Neo-Gothic engineering. The planners Raymond Hood and John Howells were chosen over numerous different engineers to structure the structure. Their Neo-Gothic structure may have spoke to the appointed authorities since it mirrored a moderate (a few pundits said "backward") approach. The veneer of the Tribune Tower is studded with rocks gathered from extraordinary structures far and wide. Other Neo-Gothic structures incorporate the Cass Gilbert plan for the Woolworth Building in New York City.

1925 to 1937 — Art Deco

Detail of high rise ventured top with needle-like top augmentation and silver ornamentation beneath

The Art Deco Chrysler Building in New York City.

With their smooth structures and ziggurat plans, Art Deco engineering grasped both the machine age and old occasions. Crisscross examples and vertical lines make emotional impact on jazz-age, Art Deco structures. Strikingly, numerous Art Deco themes were motivated by the engineering of antiquated Egypt.

The Art Deco style advanced from numerous sources. The stark states of the pioneer Bauhaus School and smoothed out styling of present day innovation joined with examples and symbols taken from the Far East, old style Greece and Rome, Africa, antiquated Egypt and the Middle East, India, and Mayan and Aztec societies.

Craftsmanship Deco structures have a large number of these highlights: cubic structures; ziggurat, terraced pyramid shapes with every story littler than the one underneath it; complex groupings of square shapes or trapezoids; groups of shading; crisscross plans like helping jolts; solid feeling of line; and the hallucination of columns.

By the 1930s, Art Deco advanced into a progressively improved style known as Streamlined Moderne, or Art Moderne. The accentuation was on smooth, bending structures and long level lines. These structures didn't include crisscross or bright plans found on before Art Deco design.

Probably the most well known craftsmanship deco structures have become visitor goals in New York City—the Empire State Building and Radio City Music Hall might be the most acclaimed. The 1930 Chrysler Building in New York City was one of the principal structures made out of hardened steel over a huge uncovered surface. The draftsman, William Van Alen, drew motivation from machine innovation for the fancy subtleties on the Chrysler Building: There are falcon hood trimmings, hubcaps, and theoretical pictures of vehicles.

1900 to Present — Modernist Styles

Smooth white flat arranged structure with focal plate formed glassed galleries

De La Warr Pavilion, 1935, Bexhill on Sea, East Sussex, United Kingdom.

The twentieth and 21st hundreds of years have seen sensational changes and surprising decent variety. Innovator styles have gone back and forth—and keep on developing. Advanced patterns incorporate Art Moderne and the Bauhaus school begat by Walter Gropius, Deconstructivism, Formalism, Brutalism, and Structuralism.

Innovation isn't simply one more style—it presents another perspective. Pioneer engineering accentuates funct

photo-1592168072506-f99808da7012.jpg

Innovation might be found in crafted by Berthold Luberkin (1901–1990), a Russian engineer who settled in London and established a gathering called Tecton. The Tecton planners had faith in applying logical, explanatory techniques to structure. Their distinct structures opposed desires and regularly appeared to challenge gravity.

The expressionistic work of the Polish-brought into the world German modeler Erich Mendelsohn (1887–1953) likewise facilitated the pioneer development. Mendelsohn and Russian-brought into the world English engineer Serge Chermayeff (1900–1996) won the opposition to structure the De La Warr Pavilion in Britain. The 1935 ocean side open corridor has been called Streamline Moderne and International, yet it unquestionably is one of the primary innovator structures to be developed and reestablished, keeping up its unique excellence throughout the years.

Innovator engineering can communicate various elaborate thoughts, including Expressionism and Structuralism. In the later many years of the twentieth century, architects opposed the sane Modernism and an assortment of Postmodern styles developed.

Innovator engineering by and large has next to zero ornamentation and is pre-assembled or has industrial facility made parts. The structure stresses work and the man-made development materials are typically glass, metal, and cement. Insightfully, present day planners oppose conventional styles. For instances of Modernism in engineering, see works by Rem Koolhaas, I.M. Pei, Le Corbusier, Philip Johnson, and Mies van der Rohe.

1972 to Present — Postmodernism

overstated current structure joining mechanical with splendid hues and components of traditional design

Postmodern Architecture at 220 Celebration Place, Celebration, Florida. Jackie Craven

A response against the Modernist approaches offered ascend to new structures that re-developed recorded subtleties and recognizable themes. Take a gander at these structural developments and you are probably going to discover thoughts that go back to old style and antiquated occasions.

Postmodern design advanced from the pioneer development, yet repudiates a considerable lot of the innovator thoughts. Joining new thoughts with conventional structures, postmodernist structures may frighten, shock, and even interest. Natural shapes and subtleties are utilized in surprising manners. Structures may fuse images to say something or essentially to please the watcher.

Philip Johnson's AT&T Headquarters is frequently refered to for instance of postmodernism. In the same way as other structures in the International Style, the high rise has a smooth, old style exterior. At the top, be that as it may, is a larger than usual "Chippendale" pediment. Johnson's plan for the Town Hall in Celebration, Florida is additionally energetically ridiculous with segments before an open structure.

Notable postmodern engineers incorporate Robert Venturi and Denise Scott Brown; Michael Graves; and the energetic Philip Johnson, known for ridiculing Modernism.

The key thoughts of Postmodernism are gone ahead in two significant books by Robert Venturi. In his noteworthy 1966 book, Complexity and Contradiction in Architecture, Venturi tested innovation and praised the blend of memorable styles in incredible urban communities, for example, Rome. Gaining from Las Vegas, captioned "The Forgotten Symbolism of Architectural Form," turned into a postmodernist exemplary when Venturi called the "revolting announcements" of the Vegas Strip insignias for another design. Distributed in 1972, the book was composed by Robert

Venturi, Steven Izenour, and Denise Scott Brown.

1997 to Present — Neo-Modernism and Parametricism

Spinning white boards encompassing dividers of glass in a ultra-current exterior

From the beginning of time, home plans have been impacted by the "engineering of the day." In the not far-removed future, as PC costs descend and development organizations change their techniques, mortgage holders and manufacturers will have the option to make phenomenal structures. Some call the present engineering Neo-Modernism. Some call it Parametricism, yet the name for PC driven plan is available to all.

How did Neo-Modernism start? Maybe with Frank Gehry's etched structures, particularly the achievement of the 1997 Guggenheim Museum in Bilbao, Spain. Perhaps it started with designers who tried different things with Binary Large Objects—BLOB engineering. Yet, you may state that freestyle configuration goes back to ancient occasions. Simply take a gander at Moshe Safdie's 2011 Marina Bay Sands Resort in Singapore: It looks simply like Stonehenge.

Renouned draftsman Frank Gehry properly places in his words which we as a whole genuinely furtively accept "Design ought to talk about its time and spot, however long for timelessness".The engineering discusses its occasions, the way of life, and the qualities and some way or another turns into the character of the period. Since initiation of humankind, everything including the exceptionally human skin has advanced and design is no special case. With time Architecture has been brought about by need, formed by vision, celebrated by pride and celebrated by legacy. On the off chance that a child conceived in the current age doesn't visit landmarks fabricated hundreds of years prior or isn't given history assignments, it will difficult for him to accept that a curve was really niched out of stone divider that the pyramid boxes he plays with are in reality undeniable structure in Egypt. That is the intensity of Transformation and drawing equal between the old and the advanced is a serious assignment.

Talking equitably the antiquated engineering and present day design are complete opposites like a whole other world with time being the skyline just harmony interfacing the two. We can't actually separate the two in light of the fact that the last is a deduction of the previous catalyzed by headway, need and obviously by the Architects themselves. So here are a couple of differentiations that will call attention to the Stark distinction between the old and the cutting edge engineering.

Utilization OF MATERIALS

One of the most obvious contrast between the old structures and the cutting edge ones is the utilization of materials. In the more seasoned occasions, stone and its subsidiaries were the fundamental or should I say the main material utilized. Regardless of whether it's the temples of medieval Europe or the Mosques in South East Asia, one can't envision these structures without stone. Then again the advanced structures, particularly the skyscraper ones can't be envisioned without a glass façade. A first level investigation of every single current structure would uncover that job of stone is restricted to that of feel. All gratitude to the progression in innovation that we presently have a plenty of choices to pick contingent upon the different standards that decide with regards to which material to utilize.

How unique is Ancient and Modern Architecture

Old Architecture-Khafre's Pyramid

How unique is Ancient and Modern Architecture

Current Architecture-The Louver Pyramid

Classification

In the event that one is to measure how significant engineering has become in our lives we have to stop for a second and glance back at the typology of the structures wherein the individuals put resources into previously and now. Back in the old occasions just strict structures like houses of worship, sanctuaries or at the most royal residences that individuals put resources into in any event as far as engineer .accordingly the antiquated design was elaborate in nature. It was essential to shroud the skeleton of the structure with pilasters, carvings, works of art and so on. While the cutting edge engineering is only very brutalize in nature. The sharp type of structures, negligible exteriors mirrors the period where we live in.

Level vs. Vertical

Visuals of the plans of antiquated structures and the cutting edge show one in number natural part of the general public absence of room. Antiquated structures were arranged marvelously, spreading on a level plane to impossible separations, truly. Subsequently the structures were spread out. Shockingly the modelers today have been denied this benefit because of the deficiency of spaces. They beat the difficulties with a progressively smaller arrangement and moving the spaces vertically. This scaled up the patterns of High ascent structures comings with its advantages and disadvantages which is an alternate discussion out and out.

How unique is Ancient and Modern Architecture

Old Architecture-Amer Palace,Jaipur

How unique is Ancient and Modern Architecture

Present day Architecture-Burj kalifa, Dubai

photo-1547349656-1a2929f7f759.jpg

Provincial TO GLOBAL

Obtusely look at two pictures beneath. On the off chance that I ask you what these structures, you would easily say that the one on the left is a congregation yet the one the privilege is... ..!!You need to conceptualize a little right? Let me tell educate you that it is a congregation also. The thing that matters is that the previous is an old church which has its customary longitudinal structure decked with curves, ringer tower and so on while the one on the privilege is a cutting edge church. With the development in design the structures today split away from the unbending nature of customs and is increasingly test in nature. You can't generally tell whether the structure is church, an exhibition hall, a theater or working of national significance. The structures have become, suppose, widespread. No big surprise, a designer from London can really structure a structure in India inspite of the social contrasts between the two areas in light of the fact that the thing that matters isn't reflected in the engineering today.

How unique is Ancient and Modern Architecture

Antiquated Architecture-Bath Abbey Church, England

How unique is Ancient and Modern Architecture

Present day Architecture-Jubilee Church - Richard Meier

On the off chance that we dive further into the investigation of this change in building wonder from Ancient to Modern, we understand it's the adjustment in human culture which has straightforwardly or in a roundabout way caused the change. It isn't generally a Change but instead move in the point of view in human needs. The development of Forts and castles halted with the world canceling the old arrangement of government and embracing different configurations of governments. The headway of designing and the creation of cement quickened the progressions to an enormous degree. We can't consider it a change yet rather a component of the procedure of advancement. With the rate at which we are advancing in all parts of life, this hole between the old and the cutting edge will expand and the outcomes might be calamitous.

Chapter-IV

Frugal Innovation

Innovation and development are made moderate and moved to bring down salary markets to take care of social issues, for example, lodging and framework. Simultaneously the arrangements are subjective, feasible and conservative. By what means would this be able to be accomplished? By lessening usefulness to what exactly is extremely essential.

At the point when organizations need to offer items for fragments with lower pay and lower buying power, they basically minimize quality to bring down expenses and costs. However, that is actually what shoppers don't need. Lower quality methods helpless usefulness and toughness.

Cheap development is something contrary to this unreasonable methodology. Target bunches are principally low-salary markets, for example rising and creating nations, yet created nations additionally have a need.

"Economical" implies that the item is diminished to those center capacities that are truly required and significant. No over-building.

These capacities and highlights arc of high calibcr, vigorous and strong.

It is planned with the goal that it is simple and easy to utilize and requires negligible upkeep.

In outline, the items are earth inviting and manageable because of structure and quality.

photo-1589828994379-7a8869c4f519.jpg

Innovation Economical Development

Parsimonious development is environmentally, financially and socially manageable.

There are immense issues to explain in the creating nations: lodging, training, drinking water, sterile, traffic ... The difficulty is that the arrangements are accessible on the planet yet there are numerous reasons why they can't move them. Enormous issues are expenses and moderateness.

The methodology of Frugal Innovation is tending to precisely these issues. Enhance for the world!

Thrifty advancement is

Natural because of high caliber and long lifetime.

Social due its capacity to illuminate significant social difficulties.

Prudent in light of the fact that it is reasonable for the lower pay markets which have a hugh potential. mainstays of practical business

Parsimonious advancement can make a high commitment to the Sustainable Development Goals (SDGs), set by the United Nations and expected to be accomplished constantly 2030.

Economical Development Goals SDGs

Model "Neulandt 3P" Our Neulandt 3P is a versatile precast plant to mass-produce solid divider and section components for reasonable lodging. It covers the parts of a Frugal Innovation in light of the fact that:

It is diminished to the most significant capacity – produce solid components in high caliber and practical through high advanced large scale manufacturing.

Powerful and enduring - it incorporates no helpless or upkeep serious innovations.

Basic – neighborhood individuals will be prepared to work the plant.

Neighborhood esteem creation through nearby sourcing and business.

It understands a significant test – the formation of reasonable lodging.

Marco and Christine in Kenya meeting partners for moderate lodging to assess the utilization of Neulandt 3P.

Model "Nokia 1100"

A well known model for a thrifty development is the Nokia 1100. The cell phone was intended for creating nations and presented in 2003. It was straightforward.

decreased to center capacities improved with significant highlights like a torchlight and a solid battery. strong and dustproof.

Economical advancement rehearses are principally connected with developing markets. This is on the grounds that thrifty development produces yields that fit developing markets' extraordinary needs and remarkable necessities — to be specific, that items and administrations must be reasonable so that poor(er) clients can bear the cost of them. As low-salary customers take an interest in the worldwide economy, their lives improve and their strengthening increments.

Thrifty development is a procedure. It is utilized to reconfigure worth and gracefully chains; it regularly applies creativity and configuration thinking practices to reproduce items and upgrade administrations. Economical advancement makes new plans of action, and it very well may be applied to create versatile tasks and maintainable strategic policies. My examination on cheap development has driven me to find three characterizing models about its training. Development procedures and practices are parsimonious on the off chance that they all the while:

Accomplish significant cost decreases

Focus on center functionalities

Add to accomplishing ideal execution

These standards apply when markets decide whether, when and how parsimonious advancement is a solid match. On the off chance that, when and how the fit is correct, parsimonious development professionals — in the case of working in organizations or advancing due to legitimate need to live — are additionally liable to practice a "cheaply creative" attitude.

Are economical trailblazers modest? Perhaps some recognize in that capacity, however I consider them to be definitely mindful of the costs in question and the normal measure of incomes to be created from selling an item or conveying a help. For organizations rehearsing economical advancement, upgraded benefits originated from a mix of diminishing (or containing) costs and expanding incomes because of selling more items or administrations in both existing and future markets.

The expression "parsimonious advancement" emerged from interpretations of the Hindi word "jugaad." The word means an "ad libbed game plan or work-around obliged by an absence of assets." The term mirrors the procedure and practice of finding minimal effort, creative answers for issues insightfully, inventively and remarkably.

I picked up bits of knowledge on how jugaad is polished in India by finding out about it from a previous understudy of mine. Niraj Lodaya is a biomedical designer instructed at the University of Florida who presently works in his local India. Lodaya refered to instances of jugaad from his expert encounters and individual perceptions: A cooler made of mud, Tata's industry-changing and society-changing Nano vehicle, the Indian government's Indian Space Research Organization (ISRO) and its Mars program.

What I gained from Lodaya was enhanced by seeing if, when and how thrifty development is rehearsed in another piece of the photo-1474631245212-32dc3c8310c6.jpg

World when I visited New Zealand organization Globex Engineering Ltd. The Auckland-based Globex has practical experience in taking care of issues through wise mechanical designing.

During the visit, one of my hosts was the organization's overseeing executive, Ed Scholten. Scholten is an accomplished and master architect, and his business theory is, "On the off chance that you can think it, we can make it." Scholten is right on target with his case, as I saw firsthand how Globex's innovative originators and gifted specialists transform imaginative thoughts into models and afterward change the models into economically fcasiblc itcms.

My involvement with Globex left me intrigued by parsimonious advancement. This is on the grounds that I saw the advantages of parsimonious advancement being drilled in a created showcase as opposed to a developing one. I was snared on the idea and needed to know more. In this way, I went to Globex's Craig Shannon. Shannon drives Globex's item configuration practice, and I asked how the organization utilizes cheap advancement when playing out its plan and building for customers.

Shannon further clarified the act of parsimonious development with a model: "One of Globex's customers represented a thorny test for our creators and architects; the customer required an on location tidy up space to fabricate their items, yet the customer didn't have the opportunity or cash to structure and construct the tidy up room they required. For us, planning and building the office was not the test; not having a lot of time and just a minimal expenditure from the customer were our group's difficulties. In this way, with a channel, an old blower, ducting, and some sheet metal, Globex creators and architects developed a smaller than expected tidy up room get together zone that diminished the customer's item get together deformities from 50 percent to 0 percent in a couple of days."

Cheap development is like different ideas that look to accomplish more by having less. The three ideas featured beneath line up with cheap development standards in that they, as well, can be influenced by constrained assets or are limited by expecting to follow a command to consistently improve results.

Worth Engineering

General Electric propelled this idea during World War II. Because of the war, deficiencies of gifted work, crude materials and segment parts were ordinary. To defeat these requirements, the organization searched for adequate substitutes and saw that the replacements it discovered regularly decreased working costs, improved a completed item's highlights and use, and upgraded laborer profitability and effectiveness. What started as a push to do whatever was important to contend in troublesome occasions transformed into an efficient procedure at first called "esteem examination."

Kaizen

As a major aspect of the Marshall Plan after WWII, American business pioneers and quality administration specialists educated and actualized strategic approaches in Japan. These practices brought about the selection of a Japanese word for development: "kaizen." Soon, Japanese organizations applied the term to portray exercises that persistently improve business activities, practices and execution.

By definition, "kaizen depends on presence of mind, self-control, request and economy." The act of constant improvement contributes essentially to accomplishing creative results; to be sure, the training is a key component found in lean assembling and creation forms. Be that as it may, nonstop improvement isn't simply restricted to assembling and creation; kaizen likewise applies to — and is generally utilized in — programming advancement and different enterprises extending from budgetary administrations to medicinal services to retail to transportation and government.

Without a moment to spare Production (JIT)/Lean Manufacturing

JIT standards are for the most part connected with Toyota's creation frameworks. The idea was resulting from need and is ascribed to Toyota's leader at that point, Fujio Cho. Cho expected that Japan's car industry would not endure exceptional rivalry except if it utilized imaginative ways to deal with get up to speed to American automobile makers. This source of inspiration added to the creation and ensuing utilization of JIT ideas, as other Japanese makers immediately received JIT creation standards and practices.

Lean assembling ideas depend on JIT basics. By expanding on the JIT establishment, lean assembling ideas place an additional attention on productivity by looking for approaches to create quantifiable incentive for clients. With this accentuation on improving client esteem, lean makers are headed to deliver something of significant worth for the client in each progression of the creation procedure.

We live during a time of worldwide commercialization. However, worldwide industrialism isn't equivalent, or reasonable or consistently spread over a planet comprising of 7+ billion individuals. The buyer pyramid is most extensive on the base, where billions of low-salary shoppers remain alive on a couple of dollars daily. But then, its a well known fact why such huge numbers of inventive items and administrations are conveyed to the a large number of high-pay customers sitting at the head of the shopper pyramid.

Worldwide commercialization follows the cash. That prompts the head of the shopper pyramid instead of the base, where billions of low-pay purchasers dwell in topographically remote nations; those in the base of thc buycr pyramid havc constraincd monctary asscts and don't approach credit. They are delegated "non-bankable" by monetary administrations organizations and are avoided by worldwide organizations.

But then, these troublesome conditions can prompt open doors for cheap trend-setters to change the lives of billions of low-pay purchasers. That is on the grounds that parsimonious advancement — in for the most part all cases — brings down the creation and circulation expenses of items and administrations without bargaining quality. Economical advancement diminishes the measure of assets expected to configuration, create, produce and offer items and administrations to the focused on commercial center — which, once more, comprises of billions of customers.

In doing as such, cheap advancement rehearses change low-salary purchasers from latent buyers into dynamic ones. What results is a worldwide economy that benefits — and benefits — from monetary development that is progressively comprehensive.

David Whitney expounds on development and enterprise and prompts organizations on everything identified with advancement and business. David is a universal speaker and has shown seminars on development and business in both school study halls and corporate meeting rooms the world over. Whitney is the Innovator-in-Residence at LeadingAgile Innovation Labs just as fills in as a business tactician at Advisory.Works. In these jobs, David applies his operational experience and topic skill while prompting LeadingAgile's and Advisory.Works' customers and vital accomplices. Whitney's direction assists organizations with accomplishing business greatness, market developments and achieve operational discoveries by applying key arranging procedures and actualizing taught execution instruments. Whitney works with organizations of all sizes on how best to collect and work innovatively disapproved, advancement arranged and benefit driven groups that more than once consider, make, and dispatch economically suitable items and administrations.

Parsimonious Innovation: How to Find Opportunity in a Storm of Adversity

Confronting impossible imperatives and affliction, numerous organizations have revealed the estimation of asset shortage as a wellspring of liberation, not hardship. An energy for extricating the most elevated conceivable incentive from existing resources – anything from aggregate information to innovation speculations – frees the whole workforce to convey outstanding arrangements. Furthermore, more regularly, their endeavors decidedly sway their workers and clients, yet in addition the general public on the loose.

This type of advancement is designated "economical development." People for the most part don't relate "thrifty" with "development." I realize I didn't, from the start, until famous planner and creator Navi Radjou clarified the subject in these terms during the Webcast, "Cheap Innovation For SAP Customers: Improve the Return on Your SAP Investment – A Webinar Event":

Cheap development is oppositely restricted to the manner in which organizations ordinarily advance. It is tied in with making increasingly monetary and social worth utilizing less assets – managing, yet improving things.

While this mentality has been around for longer than 10 years, it's reemerging in more meeting rooms. Occasions brought about by the pandemic – from sensational movements sought after to a questionable economy that may incorporate a two-year downturn – are making a direness to adjust and turn to "another ordinary" with shrewd, quick, and financially savvy arrangements.

Separating the Pharaoh complex

In this season of troublesome vulnerability, numerous organizations battle with a Pharaoh complex. They offered complex items and administrations that were sought after before the pandemic. In any case, as the market moved to other "fundamental" classes, these top of the line contributions were rendered immaterial to the requirements of cheap clients.

While numerous officials expect that the market will come back to pre-pandemic levels inside a couple of months, the truth might be entirely different, as the International Monetary Fund cautions that organizations will probably hold up until 2022.

"As the world goes into a drawn out downturn, organizations need to conquer this Pharaoh complex and have the quietude to join cost-cognizant clients where they are," exhorted Radjou. "On the off chance that the client can't go to the item or administration, the item or administration needs to go to the client."

We are as of now observing this dynamic happen over an assortment of enterprises at the present time. On location occasions, for example, seller meetings, gatherings, workshops, and in any event, counseling administrations arc currently led with remote joint effort innovation, for example, Zoom and Microsoft Teams. The equivalent is occurring in social insurance as more patients are counseled every day than any other time in recent memory through telemedicine, just as in training where understudies associate with educators and learn exercises in their lounge rooms. There's even a solid competition among researchers worldwide as they grow progressively moderate and open tests for COVID-19, extending from a $6 paper strip to a $1 unit.

Acknowledging how these administration models bring new open doors for improved commitment, decreased expenses, and included worth, organizations will naturally broaden those points of interest by scaling items and administrations. "Pushing ahead, each association will have less.

However, they should be mindful so as not to think about their representatives as lesser creatures," shared Radjou. "At the point when a restriction is put on assets, organizations expel the impediment on innovativeness."

Rearranging the earnestness to get by with resourcefulness

With a technique to endure misfortune and discover opportunity in change, organizations are exploring through remarkable vulnerability. There's no record of comparative occasions to draw exercises from nor profound stashed records to support enormous scope advancement ventures.

When confronting such an absence of assets, regardless of whether it's capital, individuals, or gear, individuals definitely have two options: gripe and surrender or ascend over the shortage with the enthusiasm to utilize what they have. They could mean utilizing internal assets like resourcefulness, compassion, and versatility; scholarly capital and information based resources; and social abilities, for example, interpersonal organizations. Or on the other hand they could be existing advancements.

"You may state, hold up a second, I don't have this a certain something. Be that as it may, possibly I have different assets that I haven't distinguished, not to mention, utilized at this point," finished up Radjou. "Whatever they might be, your assets can move motivation to commend the business' novel worth while utilizing what you as of now have."

For SAP clients, that asset could be as of now be in their current advanced scene. Watch out for future web journals in which I will plunge further into the idea of parsimonious development and SAP from various alternate points of view. In my next blog, I will investigate how you have to think as User Experience as an open door for parsimonious advancement, you can convey thrifty arrangements that can improve things significantly to your venture.

Meanwhile, you can get more bits of knowledge from Navi Radjou by tuning in to the on-request Webcast, "Parsimonious Innovation For SAP Customers: Improve the Return on Your SAP Investment

A New York-based advancement and administration mastermind who exhorts senior administrators worldwide on advancement development systems, Navi Radjou won the esteemed Thinkers50 Innovation Award in 2013 – given to an administration scholar who is reshaping the manner in which we consider and practice advancement. He conveyed a discussion at TEDGlobal 2014 on parsimonious development with over 1.8 million perspectives.

Navi co-created Frugal Innovation: How To Do Better With Less, distributed by The Economist in 2015, just as the worldwide blockbuster Jugaad Innovation with more than 250,000 duplicates sold around the world. In September 2020, he will discharge another book on how people and associations can rehash themselves intentionally to flourish in the post-COVID-19 world. He is a looked for after keynote speaker and generally cited in worldwide media .

Grasping converse and economical advancement in developing markets

Numerous officials, with an end goal to develop the business, go to developing markets, utilizing driving edge advancements to drive new market openings. While more up to date advancements are unquestionably driving development, there might be far more extensive development opportunity by utilizing existing innovations in new, inventive ways.

This development procedure is called thrifty, or switch, advancement. While these terms are some of the time utilized conversely, it merits understanding the contrasts between the two.

Cheap advancement, additionally called economical building, is the way toward diminishing the intricacy and cost of specific merchandise and their creation to make them increasingly reasonable. This for the most part includes expelling unnecessary highlights from a tough decent —, for example, a vehicle or telephone — and selling it in creating nations where shoppers can't manage the cost of the "full-include" adaptations.

Invert advancement includes growing, amazing failure cost items — in view of current advances — to sell in creating nations. The engineer at that point extends its market by selling these new items at low costs in progressively created Western nations.

Economical advancement is completely substantial. All things considered, it presents the littlest overall revenues and focuses on the most rich portions of society in these creating nations. Turn around advancement, be that as it may, appears to have the best potential for development.

In India, for instance, organizations are quickly embracing new innovations without the weight of invigorating existing expensive foundations and heritage frameworks. Ranchers in country India never had landlines; that innovation avoided an age. Presently, a rancher in India has a superior remote association than you can get in the United Kingdom.

Truth be told, in the town where I grew up, one rancher sells his vegetables by bike, riding from homestead to cultivate, however interfaces with his providers and clients utilizing his cell phone and applications, for example, WhatsApp.

In places where human services administrations are scant, organizations are sending versatile telemedicine vehicles to remote towns to permit individuals to see specialists by means of videoconferencing. These equivalent administrations are then being taken as "converse advancement" and executed in rustic regions in created markets.

As an organization official, how might you tap into this development opportunity? The key is in understanding the neighborhood networks; not all purchasers are the equivalent. Truth be told, it is basic for organizations to set up neighborhood associations so as to misuse worldwide stages however convey restricted administrations.

Different variables you ought to consider when hoping to develop in developing markets through converse or parsimonious advancement are:

Socioeconomics. For instance, more youthful advanced astute populaces will be speedier to embrace Western-enlivened advances, while buyers more than 60 will add to expanded utilization in lodging, transportation and diversion.

The rising white collar class. This gathering is developing, offering ascend to developing industrialism around the globe — remembering for developing markets.

The takeaway here is to think outside the notorious box: Look at creative ways people, governments and different organizations are taking what we may consider common innovations, for example, the web of things (IoT) and utilizing them in startling ways. Rather than looking to new advancements, expand current advances through inventive use cases. Or then again help the individuals who are discovering achievement that way, and scale to develop their business.

Globalization of the Future: How can Frugal Innovation cultivate Economic, Social and Environmental Sustainability

Worldwide force relations are moving and the dominating situation of Western economies has been essentially tested in the course of the most recent two decades by rising nations from the worldwide South. To an ever increasing extent, enormous creating economies like India, China and Brazil take part in the worldwide economy on an equivalent balance with their Western partners. Be that as it may, at the small scale level this expanded correspondence is a long way from being acknowledged and while new MNCs from the South flourish in the open market economy, the real individuals at the Bottom of the Pyramid (BoP) by and large don't pick the natural products from expanded globalization. Customary innovative work drove development and standard money related frameworks (even those utilized by Southern firms) don't accommodate their circumstance wherein buying influence is low and financial time skylines are short. Cheap Innovation, a moderately new idea in advancement the executives concentrated on the (re)design of items, administrations or frameworks to make them reasonable for low-salary clients without yielding client esteem (Peša, 2014), may conceivably be a response to unanswered BoP interest for moderate and solid correspondence, utilities, medicinal services and monetary items. It plans to bring items, administrations and frameworks inside the range of billions of poor and developing white collar class buyers at the Middle and Base of the Pyramid (Bhatti, 2012; Zeschky et al, 2014). By drastically reducing expenses while protecting client esteem and mechanical complexity, parsimonious advancement has been hailed as possibly troublesome of development forms, plans of action and even whole economies (Tiwari and Herstatt, 2012; Rao, 2013; Radjou and Prabhu, 2014).

Thrifty Innovation isn't just an intriguing new idea from the point of view of helpless purchasers. The idea of accomplishing more with less (Radjou and Prabhu, 2014) likewise offers an intriguing new way to deal with Corporate Social Responsibility (CSR), a field of studies that has gotten noteworthy consideration in the course of the most recent decades, both by and by and in principle through spotlight on issues like work codes and ccological effects and norms. Additionally in this field a discussion develops about moving viewpoints, as nations like China, India and Brazil would prefer just not to be standard taking nations, however normally have thoughts of their own on what obligation involves, impacting the idea of being a dependable business later on.

Inconsistent financial benefit of globalization is anyway by all account not the only test for good future development we face in the coming years. Ordinary monetary development has depended vigorously on normal assets and progressively it turns out to be certain that this sort of development can't be continued over the long haul. The world needs new, supportable sorts of vitality and diminishing dependence on unrenewable assets to create towards a green economy. Thrifty advancement can assume a significant job in associating the South to this turn of events and possibly even outcome in creating nations jumping towards adaptable and clean vitality and water gracefully through economical activities at the meso or small scale level. Moreover, maintainability goes past the earth and furthermore has its social and monetary viewpoints. Fruitful Frugal Innovation in a perfect world considers every one of the three components and can in this manner be a helpful apparatus in arriving at a large number of the Sustainable Development Goals (SDGs). With its attention on setting affectability, reasonableness and inexhaustibility it might impact destitution easing, comprehensiveness, and long haul supportability.

Albeit Frugal Innovation may have the ability to have any kind of effect in making manageable worldwide advancement in accordance with the SDGs, the idea has additionally been condemned as only compounding entrepreneur misuse and disparity (Schwittay, 2011; Dolan, 2012). Thusly we have to stay basic to forestall falling into the entanglement of the following publicity being developed talk. Cheap Innovation frequently happens at without a doubt the base where casual economy occupations and enterprising exercises make the greater part of (humble) wages. Too unbending change from the casual to the conventional economy through Frugal Innovation possibly undermines vocations of poor people and hence the connection between cheapness, development and the formal/casual partition should be better surveyed. Progressively definite observational examinations are required to learn whether thrifty advancement will prompt evenhanded monetary development or whether it will simply support existing imbalances (Knorringa et al, 2016).

The point of this Working Group meeting is to get a handle on how, where and when Frugal Innovation can assume a persuasive job in handling the difficulties of financially, socially and earth supportable advancement of today and our future. We welcome commitments that spread inquiries like (yet are not constrained to):

In what capacity can Frugal Innovation lead to supportable monetary change in the Global South?

In what capacity can Frugal Innovation be a main factor in the developing worldwide private division's desire to work mindfully?

Where, when and by what method can effective Frugal Innovations lead to maintainable financial development at the full scale level?

Ceaseless movement machine

Ceaseless movement is a thought which individuals from old occasions attempted to incorporate with a machine. The thought was that an appropriately fabricated machine would run always without having any consistent gracefully of vitality from outside itself. Along these lines, it would run itself while creating its own capacity to keep running.

The logical view today is that a genuine unending movement machine is a viable difficulty. The thinking is as per the following.

Any machine begins to work when provided with some unequivocal measure of vitality of the correct kind (for which the machine was structured) and of the best possible "vitality request level." High-level vitality structures incorporate those mechanical movements of a motor drive-bar or a consistent flow of power or of a hot item sending its warmth (atomic movement) into a colder article. The colder article has the most minimal request of vitality of the entire arrangement named.

As the machine fires up, the moving parts rub together and erode. Therefore they squander a portion of the first significant level movement and spread it about as worn machine sections and low-level warmth. To put it plainly, contact and warmth misfortune are the two ever-present winners of ideal utilization of vitality and in this way of ceaseless movement. The second law of thermodynamics, expresses that the warmth in a material can't be totally changed into mechanical vitality—aside from if the machine could work at supreme zero (— 460 °F., — 273.1 °C.). total zero temperature has never been reached.

The planets and regular satellites, for example, the moon, do appear to go about their focal bodies never-endingly, for they move in the close vacuum of room and experience practically no rubbing. The primary contact like powers on satellites are those made by space trash—mctcors, or comets. Such garbage—or huge estimated impacts—may at some point end the interminable movement of even these bodies.

One of the many proposed ceaseless movement machines is that portrayed above, right. It has three springs, one of which is bolstered by two upstanding poles. The other two have metal circles at their lower closes.

The framework is begun by bringing vitality into circle A; that is, it is set to vibrating all over. In the end circle B will vibrate and A will stop.

This procedure will rehash itself for a long while. At that point for what reason will it fall flat in ceaseless movement? There will consistently be some air grinding; and even in a vacuum, there will likewise be the interior erosion and warmth loss of the particles in the springs themselves. Without including increasingly outside vitality at that point, this machine will at long last overview.

Bhaskara Wheel

The absolute first know plan of a ceaseless movement machine goes back to 1150. It is known as the Bhaskara Wheel, and was planned by an Indian mathematician Bhaskara II. After some time, individuals attempted to improve the first plan — utilizing mercury, loads on enunciated arms, moving balls, and so forth.

These plans attempted to move mass to a bigger sweep from the wheel's hub, trying to have progressively mass on one side of the pivot, unbalancing the wheel to support turn.

The intriguing thing about essentially all the varieties of the Bhaskara Wheel is the means by which alluring and persuading the recommendation looks to somebody with restricted information on mechanics. The schematics of the Wheel have persuaded a considerable amount of individuals, including some well known educated people. "This should work" — was a somewhat normal response to it.

The Bhaskara Wheel is genuinely the Quintessential Perpetuum Mobile — it might be said that the first plan was reexamined again and again throughout hundreds of years, with its most recent rebirth going back to 1969. Fundamentally, at regular intervals somebody thinks of another variety, and this new thing called "the primary law of thermodynamics" doesn't appear to have impacted the recurrence of the resurrections.

Indeed, since individuals are sufficiently bold to address if the Earth is circular, it is odd to anticipate that them should be terrified of the main law of thermodynamics.

It is intriguing to take note of that Leonardo da Vinci drew various overbalanced wheels — while being for the most part against such gadgets — still was somewhat inquisitive about them.

The time somewhere in the range of 1150 and 1775, when the Royal Academy of Sciences in Paris quit tolerating recommendations "concerning unending movement", was the Wild Era of the Perpetuum Mobiles — everybody was developing and building them left and right — exhibiting them to lords and masses for a pleasant charge.

Obviously, it is guileless to expect the possibility of an unending movement machine to simply cease to exist since patent workplaces around the globe, following the French Royal Academy, quit tolerating the applications.

Besides, a few nations were progressively opened to the thought, and in 1868 Alois Drasch got a US patent for his variation of a Perpetuum Mobile dependent on "push key-type outfitting". It is difficult to state if the entirety of the

"creators" were simply preposterous or spurred by money related ravenousness during this marginally less wild, yet still turbulent time, yet some of them made millions.

One of the more known names on the rundown of innovators of ceaseless movement machines during this time is Nikola Tesla, who professed to have developed one, despite the fact that in a fairly dubious manner:

"A takeoff from known strategies — probability of a "self-acting" motor or machine, lifeless, yet competent, similar to a living being, of getting vitality from the medium — the perfect method of acquiring intention power."

In the cutting edge time the US Patent and Trademark Office (USPTO) was as yet opened to the thought:

• 1977 Emil T. Hartman gets the patent №4,215,330 for "Changeless magnet drive framework"

• 1979 Howard R. Johnson gets the patent №4,215,330 for "Electrical generator or engine structure, dynamoelectric, straight"

In 1983 USPTO at last began to get up to speed and dismissed Joseph Westley Newman's application for an interminable direct flow electrical engine he documented in 1979.

Simply joking, in 2000's USPTO allowed a patent №6,362,718 to Tom Bearden for the "unmoving electromagnetic generator", and despite the fact that the American Physical Society gave an announcement against the conceding, it feels like in the event that you have a sketch of your very own Perpetuum Mobile you despite everything can have a go at licensing it.

Karpen's Pile: A Battery That Produces Energy Continuously Since 1950 Exists in Romanian Museum

The "Dimitrie Leonida" National Technical Museum from Romania has a bizarre sort of battery. Worked by Vasile Karpen, the heap has been working continuous for a long time.

The creation can't be uncovered in light of the fact that the gallery needs more cash to purchase the security framework vital for such a show. 50 years back, the heap's designer had said it will work always, thus far it would seem that he was correct. Karpen's ceaseless movement machine presently sits made sure about right in the executive's office.

It has been designated "the uniform-temperature thermoelectric heap," and the primary model has been worked during the 1950s. In spite of the fact that it ought to have quit working decades prior, it didn't.

The Karpen heap model has been collected in 1950 and comprises of two arrangement associated electric heaps moving a little galvanometric engine. The engine moves a cutting edge that is associated with a switch. With each half revolution, the cutting edge opens the circuit and closes it at the beginning of the subsequent half. The cutting edge's turn time had been determined with the goal that the heaps have the opportunity to revive and that they can reconstruct their extremity during the time that the circuit is open.

Likewise read: $19 Device Extends Your Expensive Phone's Battery Life Like Nothing Else. 20% Discount Code: GREENOPT

The reason for the engine and the cutting edges was to show that the heaps really create power, however they're not required any longer, since current innovation permits us to gauge all the boundaries and framework every one of them in a progressively appropriate manner.

A Romanian paper, ZIUA (The Day), went to the historical center for a meeting with Nicolae Diaconescu. He removed the Karpen heap from its made sure about rack and permitted the masters to gauge its yield with a computerized multimeter. This occurred on Feb. 27, 2006, and the batteries had demonstrated a similar 1 Volt as in 1950.

They had referenced that "not at all like the exercises they instruct you in the seventh grade material science class, the Karpen Pile has one of its terminals made of gold, the other of platinum, and the electrolyte (the fluid that the two anodes are inundated in), is high-virtue sulfuric corrosive." Karpen's gadget could be scaled up to reap more force, includes Diaconescu.

The Karpen heap had been displayed in a few logical meetings in Paris, Bucharest and Bologna, Italy, where its development had been clarified generally. Analysts from the University of Brasov and the Polytechnic University of Bucharest in Romania have even performed exceptional examinations on the battery, yet didn't arrive at an unmistakable resolution.

Vasile Karpen, the innovator

"The French showed themselves exceptionally intrigued by this patrimonial article during the 70s, and needed to take it. Our historical center has had the option to keep it, however. As time passed, the way that the battery doesn't quit delivering vitality is increasingly clear, bringing forth the legend of a never-ending movement machine."

A few researchers state the gadget works by changing warm vitality into mechanical work, however Diaconescu doesn't buy in to this hypothesis.

As indicated by some who contemplated Karpen's hypothetical work, the heap he developed challenges the second rule of thermodynamics (alluding to the change of warm vitality into mechanical work), and this makes it a second-degree ceaseless movement machine. Others state it doesn't, being just a speculation to the law, and an utilization of zero point vitality.

On the off chance that Karpen was correct, and the rule is 100% right, his heap would change the entirety of the material science hypotheses from the base up, with difficult to envision outcomes. Despite the fact that I surmise this won't occur very soon, the gallery despite everything needs appropriate private subsidizing to gain the important security gear required by the police to display the gadget.

Never-ending Motion Machines: Working Against Physical Laws

Nearly when people made machines, they endeavored to make "never-ending movement machines" that chip away at their own and that work for eternity. Nonetheless, the gadgets never have and probably never will fill in as their creators trusted.

"To put it plainly, never-ending movement is inconceivable on account of what we think about the geometry of the universe," said Donald Simanek, a previous material science teacher at Lock Haven University of Pennsylvania and maker of The Museum of Unworkable Devices. "Nature gives no instances of interminable movement over the nuclear level."

Laws of Thermodynamics

As far as we could possibly know, never-ending movement machines would abuse the first and second laws of thermodynamics, Simanek disclosed to Live Science. Basically, the First Law of Thermodynamics expresses that vitality can't be made or demolished, just changed starting with one structure then onto the next. A ceaseless movement machine would need to deliver work without vitality input. The Second Law of Thermodynamics expresses that that a separated framework will push toward a condition of turmoil. Also, the more vitality is changed, the a greater amount of it is squandered. An unending movement machine would must have vitality that was rarely squandered and never pushed toward a cluttered state.

All things considered, the sacredness of the laws of material science has not halted the inquisitive from overlooking them or attempting to break them. As indicated by Simanek's online historical center, the primary archived unending movement machines incorporated a wheel made by Indian creator Bhaskara in the twelfth century. It as far as anyone knows continued turning because of an awkwardness made by compartments of mercury around its edge. Different endeavors incorporate a sixteenth century windmill, seventeenth century siphons, and a few water plants.

While most ceaseless movement endeavors have been in the soul of logical request, others have expected to trick and bring in cash. The most acclaimed never-ending movement deception was conceived by Charles Redheffer in 1812.

A period of miracles and fiendishness

Redheffer's ceaseless movement machine enchanted the Philadelphia and New York people group and acquired a large number of dollars. It was exposed twice by engineers, which at last prompted Redheffer being come up short on town, as indicated by "Unending Motion: The History of an Obsession" (Adventures Unlimited, 2015) by Arthur W.J.D. Ord-Hume.

Nineteenth-century America was a prime time for fabrications. As per Kimbrew McLeod, creator of "Pranksters: Making Mischief in the Modern World" (NYU Press, 2014), the Age of Enlightenment's attention on science, learning and picking up information through close to home understanding and perception guided expanding quantities of individuals to search out wonders that they could decide for themselves. Furthermore, expanding education rates implied that more individuals knew about ideas like ceaseless movement and were anxious to see a machine that accomplished it.

Be that as it may, as Barbara Franco wrote in "The Cardiff Giant: A Hundred Year Old Hoax," "individuals were keen on the new sciences without truly understanding them … The nineteenth century open regularly neglected to make a differentiation among well known and genuine investigations of subjects. They heard talks, went to theaters, went to interest galleries, the carnival and recovery gatherings with much a similar excitement."

Amy Reading, writer of "The Mark Inside: A Big Swindle, a Cunning Revenge, and a Small History of the Big Con" (Vintage, 2013), takes note of an exceptional trademark in the American feeling of fun. Individuals appear to appreciate being taken in by a story that they know may be false, getting bulldozed in any case and afterward being astonished after learning they were tricked. That Redheffer was really come up short on town proposes that mid 1800s crowds maybe hadn't yet completely grasped that type of diversion, however they would in ensuing decades.

Unending movement mixes Philadelphia

History specialists don't have the foggiest idea about Redheffer's experience preceding the fabrication, as per Ord-Hume. He showed up on the scene in 1812 when he opened a house close to the Schuylkill River for open survey. Inside was a machine he guaranteed could continue moving perpetually while never being contacted or in any case helped.

Redheffer's machine depended on an "expected 'rule' of ceaseless movement that accept consistent descending power on a slanted plane can deliver a constant even power segment," said Simanek. The machine had a gravity-driven pendulum with a huge level rigging on the base, as indicated by Ord-Hume. Another, littler apparatus interlocked with the bigger one. Both the huge rigging and the pole had the option to turn independently. Put on the rigging were two slopes, and on the inclines were loads. The loads should drive the huge rigging ceaselessly from the pole, and the rubbing would make the pole and apparatus turn. The turning apparatus would, thus, power the interlocked littler rigging. In the event that the loads were evacuated, the machine halted.

As per the Visual Education Project, sources contrast on the sum Redheffer charged clueless Philadelphians to see his machine. Some state he charged $5, others state he charged $1, and others state ladies were allowed in free or for $1. In any case, the cost didn't stop the entranced open, and the machine turned into a sensation. Wagers up to $10,000 were set on its realness.

Redheffer was so satisfied with his machine and its gathering that he campaigned the province of Pennsylvania for assets to assemble a bigger one. On January 21, 1813, the state sent investigators to examine before giving out the cash. It was then that Redheffer's plan self-destructed.

The primary exposing

As indicated by Ord-Hume, upon appearance, the reviewers saw that the machine was in a live with a bolted entryway and missing key. They could just view it through a window. One of the auditors, Nathan Sellers, had brought along his child, Coleman. Youthful Coleman saw that the apparatuses in the machine were not working the way Redheffer guaranteed they did. The machine gear-pieces in the riggings were worn on an inappropriate side. This implied loads, shaft, and rigging were not driving the littler apparatus aside; the littler apparatus was controlling the bigger gadget.

Nathan Sellers accepted his child and established that the machine was a deception. As opposed to defy Redheffer, in any case, he recruited Isaiah Lukens, a neighborhood engineer, to fabricate his own never-ending movement machine, which would look and "work" a similar way Redheffer's did, as indicated by Ord-Hume. Lukens built a machine that appeared as though Redheffer's yet had an apparently strong baseboard and a

square bit of glass at the top. Four wooden finials, as far as anyone knows ornamental, were on head of the glass and connected to the wooden posts. Lukens put a perfect timing engine in the baseboard. One of the finials was, actually, a winder. It could be wound and force the engine throughout the day. The engine would turn the pole, which would control the riggings.

Venders and Lukens demonstrated their machine to Redheffer, who was defeated at seeing his phony machine apparently work without a doubt, as indicated by the University of Houston's site The Engines of Our Ingenuity. He offered them cash to know how it was finished. Venders and Lukens didn't criticize him on the spot yet rather let updates on the deception spread all through the Philadelphia.

Interminable movement

Despite the fact that Philadelphia was on to Redheffer, the time's moderate correspondence speeds implied that New York was as yet an objective. Redheffer set up his machine once more. Once more, he drew huge groups. Among the spectators was Robert Fulton, a specialist most popular for building up the principal effective ad steamer. Ord-Hume composes that when Fulton saw the machine, he shouted, "Why, this is a wrench movement!"

Fulton had seen that the speed of the machine and the sound it made were lopsided, as would be the situation in the event that it were being wrenched by hand. A few reports express that the machine likewise wobbled marginally. As indicated by Ord-Hume, Fulton charged Redheffer, who raved and announced that his machine was genuine.

Fulton made an offer: Redheffer would let him attempt to uncover the genuine wellspring of the machine's vitality, and on the off chance that he wouldn't, he be able to would pay for any harm caused in the endeavor. Redheffer concurred — likely under tension from the horde of guests — and Fulton started prying off sheets from the divider close to the machine, uncovering a catgut string. The string went through the divider to the upper floor. Fulton rushed upstairs, where he found an elderly person sitting on a seat, turning a wrench with one hand and eating a covering of bread with the other.

Acknowledging they had been hoodwinked, the horde of observers devastated the machine on the spot. Redheffer fled the city right away.

Little is thought about Redheffer post-lie. As per "Resident Spectator: Art, Illusion, and Visual Perception in Early National America" (University of North Carolina Press, 2011) by Wendy Bellion, he developed another machine in 1816 however didn't let anybody see it. He was conceded a patent for it in 1820, however nothing is thought about the gadget or what happened to Redheffer. The patent itself was lost in a fire related accident.

The "difficulty" of unending movement

Redheffer's lie is history's most well known interminable movement endeavor however it is a long way from the one and only one. Most, in any case, were not intended to cheat general society out of their cash.

For what reason do individuals keep on endeavoring interminable movement machines when all laws of material science recommend they are incomprehensible?

"My hunch is that they are persuaded by their fragmented comprehension of material science," Simanek disclosed to Live Science. "The never-ending movement machine designers' perspective on material science is an assortment of irrelevant conditions for explicit purposes. They neglect to get a handle on the best quality of material science — its legitimate solidarity.

"For instance, the laws of thermodynamics don't emerge by fiat. They are resultant from Newton's laws and the active model of gases and have been all around tried tentatively ... You can't just dispose of one law you 'don't care for' without bringing the entire legitimate structure of material science smashing down."

Simanek noticed that most interminable movement machine innovators don't accept their machines abuse the laws of material science. "Some guess that specific explicit laws don't make a difference, ordinarily preservation of vitality and the laws of thermodynamics."

"Could there be some spot where the geometry (and the material science) are extraordinary?" Simanek said. "Possibly, yet we do not understand where to find that spot, and one may ponder whether we could even go there, or misuse it for our motivations ... That's easy chair theory, and sci-fi, not science."

In the event that a never-ending movement machine accomplished work, it would need to have certain attributes. It would be "frictionless and completely quiet in activity. It would radiate no warmth because of its activity, and would not transmit any radiation of any sort, for that would be lost vitality," said Simanek. All things being equal, such a machine would not run everlastingly in light of the fact that "because of its pivot, its parts would be consistently quickening, and we realize that issue is comprised of charged particles, and quickening charges transmit away vitality." This would make changes the machine, making it in the long run moderate or stop.

All things considered, "if a machine could turn a wheel at steady speed for quite a while, with no quantifiable reduction of speed, and with definitely no info vitality, we could think about it, for every single viable reason, to be interminable movement ... But it would be just a futile interest, for on the off chance that we attempted to remove work from it, it would before long delayed to a stop," Simanek said.

Most creators of never-ending movement machines have an alternate objective at the top of the priority list. "They need 'over-solidarity' execution — a machine that puts out more helpful work than its vitality input," said Simanek. At that point, you would have vitality left over for use.

Other than cheating people in general, this may have been Redheffer's definitive objective. Significantly after the lie was uncovered, Philadelphia papers estimated that the city had botched its opportunity to work water siphons for nothing, as indicated by The Engines of Our Ingenuity. Also, Redheffer's 1820 patent was for "hardware to pick up power," as per the Visual Education Project. In any case, those were wishes as opposed to real factors.

Chapter-V

Open Source Discoveries

TESLA GENERATOR

In the mid 1890s, Tesla started chip away at building up a fuel-less generator, which will deliver a wealth of vitality without the requirement for any sort of fuel.

A fruitful model was at long last evolved, and he was granted a patent for his gadget on Nov 5, 1901. He named his creation, "Device for the Utilization of Radiant Energy."

Normally known as the Tesla Generator, Tesla considered his innovation a driver of new force got from a similar vitality that controls the universe; the grandiose vitality of the sun – plentiful all over and free for all.

In spite of the fact that the seed of the generator sprouted at that point, it would take Tesla an additional 30 years of experimentation to at long last understand his vision of free vitality for all.

Vast Rays = Free Electricity

The idea driving the Tesla Generator is somewhat basic. It works on the reason of tackling the rule of leftover radio waves that are available in the climate, both on earth and in space.

To delineate how Tesla imagined vitality age from "infinite beams", let us take a gander at the unassuming precious stone radio.

The gem radio is like a Tesla Generator, in that it outfits radio waves in the environment by the utilization of a long reception apparatus made out of a wire, associated with a reasonable diode at the two finishes, with one diode "connected" aside of a high impedance headphone. The opposite finish of the headphone would in like manner be associated with a reasonable earth connector; like a water tap. Put on the headphone and what do you get? Shock, shock! You have recently made an electric-less radio player!

Going above and beyond, you can even include a variable capacitor and a loop to make a tuning circuit, so as to check out your preferred radio broadcast!

Things being what they are, does the Tesla Generator truly work?

Unquestionably! Indeed, you can produce vitality from "slim air". Be that as it may, much the same as the precious stone radio, the vitality created from a Tesla Generator is miniscule, best case scenario.

Direct Current (D/C) type electrical charge is rarely steady and loses power with separation. Besides, calling a precious stone radio a Tesla Generator resembles considering a Toyota a Lamborghini!

Outlines of a Tesla Generator

In the not all that inaccessible past, the best way to assemble a genuine Tesla Generator, was to find out about your physic reading material, become an electrical architect, and begin pouring through no under 300 distinct licenses to sift through Tesla's astuteness and find the mechanics of the Tesla Generator.

Luckily for you, our there are specialists out there that have done all the legwork for you, and through sheer difficult work, scored through piles of paper and reading material, to present to you an unmistakable bit by bit guidance manual, to construct your own special Tesla Generator now!

It won't be simple, however with real plans available to you, making your own ceaseless vitality generator is presently close enough!

The Tesla Generator – Is It Actually A Solution for Our Concerns Related To Energy?

How does the Tesla Generator furnish us with free electrical vitality?

This generator, named after the man who designed it, is a contraption that utilizes the electrical force gathered from the air to produce boundless vitality.

In view of the hypothesis that the environment has an enormous lively potential, in light of the fascination between the Earth and the Sun, which have inverse charges – one negative and the other one positive, Tesla's supporters found how to develop an attractive generator and how to utilize it to produce vitality.

After Tesla's thoughts were not, at this point a mystery, individuals that surf the Internet got some answers concerning the chance of making a generator that photo-1484100356142-db6ab6244067.jpg

Would give them boundless vitality, which they were allowed to utilize.

Albeit after the designer kicked the bucket, his thoughts were named mystery and avoided the open eye, be that as it may, some way or another, numerous decades after, these thoughts were spilled and individuals started to spread them utilizing the Internet. This is the means by which the Tesla Generator turned into a high-intrigue subject and a few people really began to consider making Tesla's fantasy conceivable.

Despite the fact that the possibility of an attractive generator that would utilize the force gathered from the air and change it into vitality that we can use as we like isn't something that individuals with fundamental information would even set out consider, building such a mechanical assembly is really something that a large portion of us can do.

You don't should be a researcher to construct your own generator; all you have to know is the manner by which to peruse. Why so? Since Tesla's thoughts have been rearranged and used to produce clear aides that anybody can comprehend and use so as to fabricate their own attractive generators. Additionally, aside from the way that you are guided bit by bit through structure up the generator, the parts you need so as to assemble it tends to be effectively found in stores. In this way, you don't require propelled information and you needn't bother with access to entangled innovation so as to manufacture an attractive generator.

Can the Tesla Generator Replace All Sources of Electrical Energy?

At the point when Tesla talked about his revelation, he said that, by building such a generator, individuals will no longer need fuel – not coal, not gas, not oil; no sort of fuel. This generator will just utilize the electrical force taken from the climate, and this will be sufficient to effectively supplant every other wellspring of power.

The Tesla Generator is one practical answer for every one of our issues identified with vitality: from its expanding cost to the harm that it brings to the earth and the way that we are going to debilitate every single common asset we discard.

This innovation probably won't be yet perceived by individuals who are skilled in this field, yet this additionally happens on the grounds that this will mean the finish of the electric organizations. Notwithstanding, things are going to change.

On the off chance that people in general becomes mindful this is a tremendous possibility for carrying on with a superior life and offering a superior future to their youngsters, the Tesla Generator will before long increase the position it merits.

German Inventor fathoms perpetual magnet engine puzzle - needs to 'part with' the disclosure...

Thomas Engel is an effective German innovator with in excess of a hundred protected creations surprisingly. He - in the same way as other of his companions - doesn't glance back at an effective school instruction, yet clearly that isn't essential for progress in the event that you are savvy and, as some state it may even be counter gainful, smothering imagination.

Engel has made sense of the working rule of a kind of engine numerous designers and hobbyists have been chipping away at - so far ineffectively. He figured out how to cause perpetual magnets to accomplish real work, changing their appealing and unpleasant force into the genuine rationale activity of rotating movement.

An ongoing article in the German every day Frankfurter Allgemeine Zeitung (12 November 2013) relates a visit of the paper's innovation editors to the designer's home and their impression of the new engine Engel says he needs to 'part with'.

While the article is deliberately composed to stay away from inconvenience and keeping in mind that it cites the required college specialists saying why such an engine is unthinkable, it gives enough detail to permit us to comprehend the idea. On the off chance that you need to begin testing, be cautioned: There is a ton of power in those uncommon earth magnets, they can be perilous to the ill-equipped.

The creator Thomas Engel shows an engine to us that never comes up short on fuel, since it works with the quality of neodymium magnets - leaving us somewhat muddled.

Specialized editors once in a while must be inconsiderate. There is a consistent stream of individuals who need to spare the world with their creations - all they are missing is open help and the money to additionally build up their thought. One needs to tell those individuals that either nobody needs their creation or that it won't work.

In any case, imagine a scenario where it's diverse this time. There is a man who is trying to reach us. He is experiencing an independent picture taker and says he has an engine at his home that has been running since April immediately and that needs no fuel to do that. The picture taker has seen it and is all eager. We have found out about such engines previously, however have never observed a show. Ordinarily, you wouldn't need to try to peruse past this point, since something like this can't genuinely exist. In any case, this time we're not discussing some wrench with a thought. This one was respected in 1972 with the renowned Rudolf Diesel Medal for creators, he has well over a hundred licenses in his possession and has been addressing at colleges everywhere throughout the world.

In the nineteen-fifties Engel built up another strategy for the creation of polyethylene making plastic funnels impervious to heated water. The Munich Olympia arena has a garden warming framework dependent on this innovation. He turned into a tycoon before the age of 30. He never at any point took the A-level school test as in 1944 he was drafted into military air protection. From that point onward, he didn't have time any more. Our diary detailed finally about his carrcer as a diswashcr prcciscly 13 ycars back (on November 22, 2000). That isn't the life story of a scoundrel. The designer lives in Baden-Baden and he has a spot in Lucerne, Switzerland, where the engine is found. So we drive off with blended emotions to lovely Switzerland.

Engel's engine is running. During the three hours we are there, it is chugging along discreetly, interfered with just by certain tests we will discuss later. There is no perceptible improvement of warmth. The appears to be recognizable, the engine acquires its capacity from neodymium (NdFeB) magnets. Those are the most grounded changeless magnets referred to, a plate as meager as a one-Euro coin can hold about kilograms of weight. Neodymium is an uncommon earth component, much utilized in gadgets. Magnets made out of this material are utilized in atomic turn tomography and in wind generators, they drive water siphons of substantial trucks and keep instruments consistent.

The magnets are produced utilizing a blend of neodymium, iron and boron which is squeezed into structure and sintered. They are then polarized with a solid electric drive. The vitality utilized for polarization, notwithstanding, isn't what keeps the magnet working. A few providers of those magnets have guaranteed us that the intensity of the magnets doesn't reduce - much following quite a while of utilization. So it appears that the magnets can accomplish work continually without getting corrupted. The main thing those magnets don't care for is incredible warmth.

Engel's thought was that it should be conceivable to change over that intensity of the magnets into rotational movement. He constructed a machine made of metal, looking like a smaller than expected machine. The rotor is a circle with magnets fixed to it. The pole turns in fired orientation. A circle magnet fixed at the right point and good ways from the rotor however which itself can turn (Engel considers it the mirror) can influence the rotor magnets. There is appealing and horrible power, contingent upon the direction of the posts: the rotor would thus be able to be set in consistent movement, as long as the mirror continues pivoting. The mirror's turn controls the speed of the rotor.

The creator and his engines. The more established rendition on the left was developed from an old watch creator's machine.

The specific structure and demeanor of the parts is hard to determine, Engel needed to try finally with those boundaries. In the event that the mirror is a touch excessively removed, the attractive field separates. Then again, in the event that it is excessively close, the neodymium magnets will tear the development separated. The mirror hangs in a sort of outrigger. Two electric wires associate with the lower end with crocodile cuts. There is a minuscule electric engine that turns the mirror. So it is preposterous to expect to manage without power by and large? The designer flags his difference. "Eight milliamperes at nine volts", he says. That is just a control instrument. The force at the pole is a lot more noteworthy. Engel additionally pondered a mechanical drive for the mirror legitimately from the rotor shaft, however selected against this as it would extensively increment mechanical multifaceted nature.

We needed to know more. The revolution is around 400 RPM. We don't have an instrument to gauge mechanical force. So we are utilizing the finger brake. It is hard to stop the revolution by getting the pole. The engine just grinds to a halt after significant warmth created on the calluses of our hands. A little hand made propeller out of plexiglass doesn't dazzle the engine by any means; we might truly want to know how much force the machine turns out. With a touch of smoothness, one can turn the mirror by hand and set the rotor moving. There is not really any obstruction when turning the mirror. We thusly danger an attestation: The yield felt at the pole is obviously more prominent than the information expected to give the motivation. Obviously estimation was just finished with human sensors.

It is conceivable to put a second rotor on the contrary side, to be tended to by a similar mirror. Holding a screwdriver between the mirror and the rotor in activity brings about a wavering movement of the screwdriver between the magnets, without anyway contacting them. Mr. Engel might want to accomplish more experimentation with the quantity of magnets and their structure, however he says he comes up short on the quality for additional turn of events.

Science is incredulous. An engine which creates more vitality than it goes through is outlandish, says Markus Münzenberg, an educator for exploratory material science at Göttingen University. Since in shut frameworks, the entirety of vitality is consistently equivalent. The evidently high force at the pole could be an outcome of inertial mass of the machine which, once moving, is hard to stop.

Ludwig Schultz, the chief of the Institute for Metallic Materials in Dresden concurs. While it is conceivable to envision magnet arrangements that intermittently pull in and afterward repulse different magnets, yet all things considered the potential vitality would occasionally seep off without there being an addition in vitality.

The innovator's response to the inquiry whether his engine is a ceaseless movement machine is to some degree angry: "That is trash", he says. "There is nothing of the sort". Mr. Engel is persuaded that his machine utilizes the colossal vitality which is natural in quanta, those incomprehensibly little segments of molecules which were first depicted by the physicist Max Planck in the early piece of the only remaining century. He in this way considers his machine a "quantum deviation device". Somethings are as yet hazy, additionally for the innovator himself. Some place in Germany, a specialist has a second such engine at his organization, which runs with 1200 RPM. The man called a few days back he says, and related that, when the engine was secured with an acrylic hood, its rotational speed decreased.

Engel doesn't have the foggiest idea about the purpose behind this.

The articulation "quantum engine" carries some terrible relationship with it, since certain cheats, about 10 years back, utilized that name to gather cash for a machine which never appeared. Engel's engine is very not the same as that, aside from a likeness in the portrayal. The designer needn't bother with cash. He says he needs to part with the engine since humankind needs moderate vitality. It must be additionally evolved until certain years later, we will make power with it in the storm cellars of our lodging units.

What is the subsequent stage now? Engel needs to join a little generator to the pole and show that his engine conveys more power than is required for its control. In the event that he could do that, we'd truly have some thrilling news.

Others are chipping away at changeless magnet engines. Here is a case of the rule that Engel utilized in another, free turn of events

Zero-point vitality (ZPE)

It is the least conceivable vitality that a quantum mechanical framework may have. Dissimilar to in old style mechanics, quantum frameworks continually vacillate in their least vitality state as portrayed by the Heisenberg vulnerability principle.[1] As well as particles and atoms, the vacant space of the vacuum has these properties. As indicated by quantum field hypothesis, the universe can be thought of not as disengaged particles however persistent fluctuating fields: matter fields, whose quanta are fermions (i.e., leptons and quarks), and power handle, whose quanta are bosons (e.g., photons and gluons). Every one of these fields have zero-point energy.[2] These fluctuating zero-point fields lead to a sort of reintroduction of an aether in physics,[1][3] since certain frameworks can distinguish the presence of this vitality; in any case, this aether can't be thought of as a physical medium on the off chance that it is to be Lorentz invariant to such an extent that there is no inconsistency with Einstein's hypothesis of extraordinary relativity.

lthough indicated by creative mind, it wasn't until 1927 when Physicist Werner Heisenberg discovered that the (a) development and (b) precise situation of particles stay unsure because of the difficulty of precisely disconnecting the two factors at the same time. In an article distributed by the Scientific American, Phillip Yam composes that if a framework was stale totally at total zero, the two factors (an) and (b) would be precisely known. Heisenberg's law approved the presence of endless and imperceptible vitality, making computing the two factors unthinkable and figuratively dropkicking ideas of "void space".

Increasingly solid proof surfaced in 1947 when Hendrik Casimir found that in the hole between impeccably directing equal plates, portions of the electromagnetic range are rejected. These limited frequencies make a weight outside the plates that goes about as a power pushing them internal. This is known as the Casimir Effect.

Zero-point vitality can be seen by the consistent fluid province of Helium, which stays steady paying little heed to barometrical temperature. Truly, zero-point vitality can be felt while circumventing a bend because of the opposition of inactivity.

Physicist Robert Forward estimated real use of this revelation in 1984. His hypothesis concerned the change of zero-point vitality into usable electric vitality by means of the development of a "vacuum variance battery" in view of the conductive plate structure of the Casimir Force. Forward expressed that permitting the power to rise above—along these lines pushing the plates to contact—would make enough vitality to change zero-guide vitality toward electric vitality.

Casimir Impact

The ramifications of zero-point vitality would be change in outlook in a worldwide economy inclining toward limited fills and the industrialist standards of gracefully and request. It is for this equivalent explanation that the course of events of logical advancement may have been so moderate—if a zero-point generator were assembled, it would be a one-time speculation.

A significant complaint by some science networks is the second law of thermodynamics. Nobel Prize champ, Ilya Prigogine has indicated that the second law of thermodynamics can be extended to incorporate non-harmony frameworks that develop to expanding request. This development makes the second law good—in principle.

photo-1470137237906-d8a4f71e1966.jpg

The possible use of zero-point vitality by the general population, should it be permitted to enter the worldwide commercial center, would fill in as an expected answer for probably the best current issues confronting mankind. Utilizing zero-point vitality would mean a perfect and supportable other option, decreasing the atmosphere changing impacts of an Earth-wide temperature boost.

The possibility of vast vitality created on a mass scale could work to democratize transportation and electric innovation in a social scene troubled by financial uniqueness, and connectedly, the computerized separate. This could battle the social original of the "have" and the "have not".

Outside our own environment, boundless vitality infers an excessive chance of investigation in a universe that is continually growing because of vacuum variances.

In a progressively powerful ramifications, the cyclic and lasting nature of vitality may give further and all the more experimentally solid understanding into development of life-power vitality after death and into the connection among development and awareness.

The free-vitality activity turns out to be progressively significant considering intensifying situation, as zero-point vitality presents us with the capability of opportunity from innovation affixed to limited asset. At the point when the regular system of buyer driven science has harmed prospect, unique logical pathways must turn into the new wilderness. The zero-point vitality activity should turn into an increasingly focal concentration in scientific networks should humankind wish to make sure about its questionable future and satisfy its pretentious and apparently inborn motivation to accomplish god-such as self-independence.

Material science at present comes up short on a full hypothetical model for understanding zero-point vitality; specifically, the disparity among conjectured and watched vacuum vitality is a wellspring of significant contention.Physicists Richard Feynman and John Wheeler determined the zero-guide radiation of the vacuum toward be a significant degree more noteworthy than atomic vitality, with a solitary light containing enough vitality to heat up all the world's seas. However as per Einstein's hypothesis of general relativity any such vitality would float and the test proof from both the extension of the universe, dull vitality and the Casimir impact demonstrates any such vitality to be especially powerless. A mainstream recommendation that endeavors to deliver this issue is to state that the fermion field has a negative zero-point vitality while the boson field has positive zero-point vitality and in this manner these energies some way or another counterbalance one another. This thought would be valid if supersymmetry were a definite evenness of nature; nonetheless, the LHC at CERN has so far found no proof to help supersymmetry; in addition, it is realized that if supersymmetry is substantial by any means, it is all things considered a wrecked balance, just obvious at high energies, and nobody has had the option to show a hypothesis where zero-point abrogations happen in the low vitality universe we watch today. This error is known as the cosmological consistent issue and it is one of the best unsolved puzzles in material science. Numerous physicists accept that "the vacuum holds the way in to a full comprehension of nature

The Zero Point Energy (ZPE) is a characteristic and unavoidable piece of quantum material science. The ZPE has been considered, both hypothetically and tentatively, since the disclosure of quantum mechanics during the 1920s and there can be no uncertainty that the ZPE is a genuine physical impact. The "vacuum vitality" is a particular case of ZPE which has created impressive uncertainty and disarray. In a totally vacant level universe, estimations of the vacuum vitality yield boundless estimations of both positive and negative sign- - something that clearly doesn't relate to the idea of this present reality.

Perception shows that in our universe the fabulous all out vacuum vitality is amazingly little and conceivably precisely zero. Numerous scholars presume that the all out vacuum vitality is actually zero.

It certainly is conceivable to control the vacuum vitality. Any items that change the vacuum vitality (electrical transmitters, dielectrics and gravitational fields, for example) misshape the quantum mechanical vacuum state. These adjustments in the vacuum vitality are regularly simpler to ascertain than the complete vacuum vitality itself. At times we can even gauge these adjustments in the vacuum vitality in research facility tests.

In traditional material science, in the event that you have a molecule that is followed up on by some preservationist power, the absolute vitality is $E = (1/2) mv^2 + V(x)$. To locate the traditional ground state, set the speed to zero to limit the motor vitality, $(1/2)m\ v^2$, and put the molecule at where it has the most reduced potential vitality $V(x)$. In any case, this outcome is just an old style guess to this present reality. Since the old style ground state totally determines both the molecule's speed (zero) and position (at the base), it disregards the acclaimed Heisenberg Uncertainty Principle ($m\ dv\ dx > hbar$). Quantum material science, by means of the Uncertainty Principle, powers the molecule to spread out both in position and speed thus makes it have a vitality to some degree higher than the old style least. The ZPE is characterized as this move:

$E(ZPE) = E(\text{quantum least}) - E(\text{classical least}) > 0$

Traditionally, we can ascertain the normal swaying recurrence that the molecule would have if we somehow managed to give it a little push. Quantum precisely, it is presently an undergrad exercise to utilize the Heisenberg vulnerability connection (all the more definitely, Schroedinger's differential condition) to show that

$E(ZPE)$ approx. $= (1/2)\ hbar\ omega0$

where hbar is Planck's steady occasions and omega0 is the normal swaying recurrence. The ZPE in this sense shows up all over the place: it influences atomic bonds, dense issue material science, little motions of any framework.

The following stage is to understand that the electromagnetic field can be thought of as an endless assortment of coupled oscillators- - one at each point in space. Once more, the old style ground state is the situation wherein the electric and attractive fields both must be zero. Quantum impacts imply that this case doesn't remain constant; there is additionally a Heisenberg vulnerability standard for electric and attractive fields (it's somewhat more mind boggling). Fortunately the potential for electromagnetism is actually quadratic thus can be unraveled precisely. The terrible news is that there is a vast number of modes. Officially we can compose

Notice

(Electromagnetic vacuum vitality) = total over all modes (1/2) hbar omega(mode)

The boundlessness in this condition is what empowers the free lunch swarm (the forefront family members of the endless development swarm), who envision a wearisome ZPE for humankind to exploit. Not actually, unfortunately..

The first and most clear issue is that there are other quantum fields known to man isolated from electromagnetism. Electrons, for one thing, notwithstanding neutrinos, quarks, gluons, W, Z, Higgs, and so on. In particular, if you do the mean electrons you will find that what are known as Fermi estimations offer rising to an extra short sign in the figuring.

Adding short unlimited quality to notwithstanding endlessness gives mathematicians awful dreams and even makes theoretical physicists stress a piece. Fortunately, nature doesn't worry over what the mathematicians or physicists think and completes the obligation regarding us normally. Consider the staggering total vacuum imperativeness (when we have incorporated all quantum handle, all atom interchanges, proceeded with everything restricted by catch or by criminal, and taken all the most ideal cutoff focuses continuously end). This extraordinary hard and fast vacuum imperativeness has another name: it is known as the "cosmological steady," and it is something that we can evaluate observationally.

In its one of a kind sign, the cosmological steady was something that Einstein put into General Relativity (his theory of gravity) by hand. Particle physicists have since accepted power over this idea and appropriated it for their own by giving it this inexorably physical delineation to the extent the ZPE and the vacuum essentialness. Astrophysicists are right now clamoring setting observational tops for the cosmological consistent. From the cosmological viewpoint these cutoff focuses are still really wide: the cosmological consistent may offer up to 60 percent to 80 percent of the full scale mass of the universe.

From a particle material science viewpoint, regardless, these cutoff focuses are exceptionally extreme: the cosmological reliable is more than $10(-123)$ times humbler than one would guilelessly evaluate from atom physical science conditions. The cosmological steady could possibly be really zero. (Physicists are up 'til now battling on this point.) Even if the cosmological consistent isn't zero it is emphatically little on an atom material science scale, little on a human-planning scale, and too microscopic to even think about being in any capacity any possible wellspring of imperativeness for human needs- - not that we have any savvy considerations on the most ideal approach to accomplish tremendous degree controls of the cosmological predictable regardless.

Putting the more interesting dreams of the free lunch swarm aside, is there much else possible that we could use the ZPE for? For no good reason, little degree controls of the ZPE are in all actuality possible. By introducing a channel or a dielectric, one can impact the electromagnetic field and along these lines instigate changes in the quantum mechanical vacuum, provoking changes in the ZPE. This is what underlies an unpredictable physical miracle called the Casimir sway. In an old style world, absolutely unprejudiced conductors don't pull in one another. In a quantum world, in any case, the fair-minded conductors upset the quantum electromagnetic vacuum and produce constrained quantifiable changes in the essentialness as the conductors move around. Every so often we can even learn the modification in imperativeness and complexity it and examination. These effects are to a great extent no ifs, ands or buts authentic and uncontroversial yet little.

photo-1514580426463-fd77dc4d0672.jpg

Progressively questionable is the suggestion, made by the physicist Julian Schwinger, that the ZPE in dielectrics has something to do with sonoluminescence. The jury is still out on this one and there is a lot of amicable discussion going on (both among experimentalists, who are dubious of which of the fighting parts is the correct one, and among researchers, who regardless of everything contrast on the specific size and nature of the Casimir sway in dielectrics.) Even dynamically hypothetical is the suggestion that relates the Casimir effect on "starquakes" on neutron stars and to gamma bar impacts.

In overview, there is no vulnerability that the ZPE, vacuum essentialness and Casimir sway are really certifiable. Our ability to control these sums is confined but at this point and again precisely interesting. However, the free-lunch swarm has gigantically distorted the importance of the ZPE. Considerations of mining the ZPE should along these lines be treated with unbelievable doubt

From the way where a couple of fans talk about the zero-point imperativeness, one may accept that endless power is lying all around essentially holding on to be outfitted. Similarly as different contemplations that have all the earmarks of being ridiculous, this one falls to pieces on closer evaluation, regardless of the way that the possibility of the zero-point essentialness is charming without anyone else. John Obienin, a materials science researcher at the University of Nebraska at Omaha, explains:

"Zero-point imperativeness implies unpredictable quantum changes of the electromagnetic (and other) power handle that are accessible any place in the vacuum; by the day's end, an 'empty' vacuum is actually a seething cauldron of essentialness. This essentialness is accessible even at absolute zero temperature (- 273 Celsius),and clearly, regardless, when paying little mind to is accessible. The effect of these vacuum fields has been recognized imperceptibly - the effect is little - by the interest they start in a capacitor, which is incredibly just two close equivalent metal plates. This effect is the praised desire for Hendrick B. G. Casimir (made in 1948); it was generally 'confirmed' likely by M. J. Sparnaay in 1958. A progressing, for the most part noted examination by Steven K. Lamoreaux (Physical Review Letters, Vol. 78, No.1, pages. 5-8; January 6, 1997) gave an amazingly accurate and unambiguous certification of the nearness of the Casimir power.

"These vacuum instabilities may have impacts, both unnoticeable and gross, on the direct of minuscule particles and on our general environmental factors. Russian physicist Andrei Sakharov hypothesized that they may offer rising to the intensity of gravity. At present, nobody acknowledges how to abuse the zero-point imperativeness in a detectable device that passes on sizable proportions of essentialness. There is, in any case, a great fringe segment (like those pulled in to UFOs, soothsaying, numerology, and so forth) of people who speculate and fantasize about the opportunity of abusing the zero-manage essentialness toward achieve diverse specific marvels and the since a long time back searched for 'wearisome development.' Consider yourself advised."

John Baez is a person from the math staff at the University of California at Riverside and one of the referees of the on-line sci.physics.research newsgroup. He incorporates some particular condition:

"Vacuum essentialness shows up in explicit computations in quantum field theory, which is the gadget we use to coordinate present day atom material science. Truth be told, particles interface with one another through a variety of forces. This is a tangled business, so in quantum-field speculation we start by analyzing a romanticized model in which particles don't associate in any way shape or form. This is known as a 'free-field speculation.' Then we use this free-field theory as the explanation behind considering the 'interfacing field theory's we are really roused by.

"In quantum-field speculation, the vacuum state is described to be the state having the least essentialness thickness. Something intriguing happens when we use a free-field theory to consider a partner field speculation: the vacuum state of the free-field theory isn't equivalent to vacuum state of the interfacing field theory. The vacuum state of the teaming up field theory may have essentially imperativeness than that of the free-field speculation; what makes a difference is known as the vacuum essentialness.

"One should not take this vacuum essentialness too really, regardless, considering the way that the free-field speculation is just a numerical instrument to empower us to appreciate what we are genuinely enthused about: the partner theory. Simply the partner speculation ought to relate clearly to this present reality. Since the vacuum state of the conveying theory is the state of least essentialness in reality, it is incredibly far-fetched to expel the vacuum imperativeness and use it for anything.

"It is to some degree like this: state a bank felt that it was continuously useful (strangely) to start counting at 1,000, so that regardless, when you had no money tucked neatly away, your record read $1,000. You may get invigorated and endeavor to spend this $1,000, yet the bank would state, 'Sorry, that $1,000 is just a relic of how we do our bookkeeping: you're as a general rule absolutely done for.'

"So likewise, one should not get one's desire up when people talk about vacuum imperativeness. It is actually how we do our bookkeeping in quantum field theory. There is impressively more to state concerning why we do our bookkeeping this sharp way, yet I will stop here."

Paul A. Deck, associate instructor of science at Virginia Polytechnic Institute and State University, gives a manufactured perspective on this request:

"The zero-point imperativeness can't be harnessed in the traditional sense. Zero-point imperativeness is that there is a restricted, least proportion of development (even more exactly, engine essentialness) in all issue, even at inside and out zero. For example, engineered protections continue vibrating in obvious habits. However, releasing the imperativeness of this development is unfathomable, considering the way that then the molecule would be left with not actually the base entirety that the laws of quantum material science anticipate that it should have."

CHAPTER SEVEN

Chapter-VI

Greatest Inventions of All Time

World has witnessed some greatest inventions. Which has amazingly changed the world. Here are some followings.

Remote trade of power:

This idea was one new of his various contemplations and at his time was excused. However, by and by, 70 years after his passing, it is another advancement that is starting at now used in such applications as remote charging of batteries or batteries of electric vehicles or transports. Tesla expected to transmit messages, correspondence and even duplicate pictures over the Atlantic to England and to ships afloat reliant on his theories of using the Earth to lead the signs. In case the undertaking worked, anyone could have power by basically remaining a rode into the ground. Grievously, free force isn't beneficial. Moreover, this structure could be incomprehensibly perilous for the overall top notch since it could essentially change the imperativeness business. Imagine how exceptional the world would be if society didn't require oil and coal to work? Could the unprecedented world powers care for control? Morgan wouldn't finance the changes. The endeavor was given up in 1906 and never got operational

Thought camera:

This idea was first pronounced by Tesla in 1883 with the purpose of making a device for shooting the human cerebrum. He suggested that any thought in the human cerebrum outlines an image in the retina that can be caught by a genuine device.

He never winning to carry his theory into this present reality, anyway today analysts have made sense of how to structure a phony retina using an obfuscated logical figuring reliant on which eye's retina can change over pictures into electrical signs that are sent to the psyche for various techniques.

Chronovision

Father Ernetti is interesting not considering his work as an exorcist in the Venice region, anyway more especially because of his work on the "chronovision". During the 1960s he is said to have ensured he built up a period watcher of sorts during the 1950s, as a component of a social occasion that to the extent anybody knows included Nobel Laureate Enrico Fermi and Wernher von Braun. The machine was known as the Chronovisor, and could purportedly watch and hear events of the past. As demonstrated by an explanation by Ernetti, the glowing imperativeness and sound that articles transmit are recorded in their condition, with the ultimate objective that fitting use of the chronovisor could imitate from said essentialness the photos and traces of a specific plan of events from a prior time. Through the overview screen of the chronovisor Father Ernetti proclaimed to have seen an introduction in Rome in 169 BC of the now-lost disaster, Thyestes, by the father of Latin stanza, Quintus Ennius. He also proclaimed to have seen Christ passing on the cross. On his passing bed in 1994, Father Ernetti said that he went to a get-together of the significant number of people related with the chronovision at the Vatican during which the fundamental existing machine was crushed.

Shake Machine

At one point while investigating various roads in regards to mechanical oscillators, Nikola Tesla probably made a resonation of a couple of structures making complaints the police. As the speed created he hit the resonation repeat of his own structure and belatedly understanding the danger he needed to apply a sledge hammer to end the investigation, correspondingly as the shocked police appeared. The Discovery Channel's standard MythBusters show investigated Tesla's case that he had made a "Seismic tremor Machine" in their 60th scene. They attempted the physical wonder known as mechanical resonation on a traffic interface, which today are attempted to withstand such powers. While a lone I-light emanation was diverted a couple of feet toward each way by their oscillator, and they as far as anyone knows felt the augmentation shaking various yards away, there were no "earth shattering" impacts. It justifies exhibiting that, in the hour of the event grasped by Tesla, structures were not attempted to withstand such resonation.

Loathsome power Device

In 1956, the flight trade dispersion Interavia itemized that Thomas Townsend Brown had increased liberal ground in frightful power or electrogravitic stimulus research. Top U.S. avionics associations had in like manner gotten related with such investigation which may have gotten a masterminded subject by 1957. In spite of the way that the effect he found has been exhibited to exist by various others, Brown's work was flawed in light of the fact that others and even he himself acknowledged that this effect could explain the nearness and movement of unidentified flying things (UFOs). Gritty hued's assessment has since become something of a notable enthusiasm far and wide, with beginner experimenters mirroring his underlying examinations as "lifters" powered by high-voltage.

Ozone Therapy

A couple of individuals, including different masters and natural physicists, acknowledge ozone has excellent patching properties. The presence of careful clinical ozone generators has starting late allowed the instruments, movement and possible hurtfulness of ozone to be surveyed by clinical fundamentals. Regardless, despite story confirmation of ozone treatment having caused decrease in a grouping of diseases, healing use of ozone isn't upheld by prosperity masters or clinical relationship in any English talking country, and most US states disallow the promoting of ozone generators, its clinical use, and even investigation and clinical starters of ozone treatment, with the objective that authorities chance losing their clinical licenses by overseeing or underwriting ozone medicines.

End Ray

Nikola Tesla proclaimed to have built up a "passing shaft" which he called Teleforce during the 1930s. The device was fit for creating a genuine concentrated on light discharge "that could be used to dispose of foe warplanes, outside military, or whatever else you'd ideally didn't exist". The indicated "passing pillar" was never evolved considering the way that he believed things being what they are to be exorbitantly basic for territories to beat each other. Tesla suggested that a nation could "crush anything moving closer inside 200 miles... [and] will give a mass of power" in order to "make any country, immense or little, secure against military, planes, and various techniques for attack". He said that attempts had been made to take the creation. His room had been entered and his papers had been examined, yet the hoodlums, or spies, left with scarcely a penny.

Cloudbuster

There is decidedly something to be said about the ability to make it deluge on hand. Wilhelm Reich, an analyst who saw a drought that was influencing the blueberry accumulate in his region of Maine, made a creation that has since been named "the Cloudbuster."Sound unnecessarily much like science fiction? According to the Bangor Daily News, which was covering the essential undertaking with this machine, there was no guess for any storm in the area. Inside significant lots of Reich setting completely operational the machine, storm fogs molded and accomplished 0.64 centimeters (0.25 in) of rain.It seems like Reich's development sabotaged some establishment inside the organization as his investigation was shut down and his work and models were seized. There was never a second testing of the Cloudbuster machine. Regardless, if it had been made, food insufficiencies may stop to exist the world over.

Nuclear Energy For Residential Use

Nuclear essentialness used to be an outstandingly discussed strategies for conveying colossal volumes of ability to neighborhoods everywhere throughout the world. With the ability to furnish a little division of nuclear power, entire systems could have had power for a clearly ceaseless proportion of time.When this development was close to the very edge of showing up at an alluring time of creation, theorists suddenly lost interest and the advancement became mixed up in the specialties and hole of some clammy office. The structure for this development was a little nursery type shed that would be halfway arranged in neighborhoods.This would have been a power place for the entire zone, furnishing a couple of squares of occupants with power simultaneously. Best of all, it ought to be a free or simplicity answer for vast power, which would have saved everyone from their stream high force bills.

Sloot Digital Coding

This development has just truly been "absent" since 1999. With the progressions in current innovation, this coding creation could have really reformed the space and capacity abilities of the advanced PC and versatile device.The Dutch engineer's name was Romke Jan Bernhard Sloot. With his tech, information could be consolidated altogether. The outline used to both test and market this innovation was the pressure of a full-length film to 8 kilobytes in size.The exacting calculation for the disentangling procedure was an insignificant 370 megabytes. Sloot had the option to show the accomplishment of his task by all the while playing 16 full movies from a solitary 64-kilobyte chip. As purchasers and financial specialists arranged for this mind boggling creation, Sloot kicked the bucket under dubious conditions only days before he should hand over the first source codes.

Completely Electric Car (Non-Hybrid)

In the late 1990s, GM was the first to discharge and market a completely electric vehicle. While this probably won't appear to be particularly noteworthy with the high volume of half breed vehicles that presently exist, this vehicle was the first of its kind.Even with the present progressively liberal methodology in joining gas motors with those that can work with simply electric force, there still can't seem to be an all around advertised vehicle that requires no fuel by any means. The GM EV1 was not proposed to be excessively effective; GM just made 800 of them to start.However,

as indicated by reports from the time, GM accepted that clients were disappointed with battery power and concluded that they would scrap the whole line of vehicles in lieu of further developed gas fueled alternatives out there. There is a conviction that despite the fact that GM would have sold an enormous portion of the electric-controlled vehicles, their actual rationale in rejecting their arrangement was serious weight from large oil organizations.

Solution For Heart DiseaseWith coronary illness being a main executioner among ladies the world over (and a lot of men also), realizing that a doable remedy for the condition once existed is somewhat disturbing. Much like the Rife machine recorded later, this is another development that was smothered in light of the fact that it conflicted with the momentum treatment routine for coronary illness at the time.According to the American Medical Association (AMA), this would dishonor the treatment itself as well as the specialists who were behind it. Their profession transparently challenged the way that there were recorded instances of coronary illness being relieved through this "Bound together Theory of Human Cardiovascular Disease."However, some heart patients who attempted the treatment narratively announced an intensifying of their conditions.Hemp BiofuelCommonly confused with cannabis, hemp has consistently had a terrible notoriety with the individuals who don't comprehend its actual advantages. At the point when you separate ethanol in wealth from this ground-breaking plant, it shows the genuine value.Of course, with the conviction that hemp is related with cannabis, the main hotspot for ethanol right currently is corn. Notwithstanding, the hemp plant can create more ethanol than corn and it is far less harming to the earth all the while.

4The Ogle Carburetor

Everybody might want to get more mileage per top off, and a portion of the later mixture vehicles have had the option to work superbly of getting you farther with less gas. Sadly, we still can't seem to achieve anything very like the recorded accomplishments of technician Tom Ogle.In the 1970s, this designer made another kind of carburetor, any semblance of which had never been seen. Much like today, gas and oil organizations in those days had a syndication in the market. In spite of the fact that Ogle's carburetor was tried and appeared to make a trip as much as 48 kilometers for every liter (113 mpg), his creation was never delivered commercially.The progressive segment worked by pressurizing fuel into a fume cloud, which was then infused into the terminating chambers. Permitting mishaps and obstacles guaranteed that the carburetor was never mass-delivered for use in vehicles, and Ogle kicked the bucket with its plan data.

Overflowing Device

In 1934, Royal Rife made a machine to impact away malignant growth. At that point, disease was still regarded to be an infection. So Rife made a laser light emission to focus on the particular contaminated cells and wipe out them.Think it's a fabrication?According to The Cancer Cure That Worked: 50 Years of Suppression by Barry Lynes, 14 recorded instances of terminal malignant growth patients being restored with this treatment may persuade you in any case. In any case, when Rife wouldn't join forces with the top of the AMA, the association utilized their full weight and assets to limit and dishonor the treatment.Now there is clearly no documentation to demonstrate authoritatively that the AMA had such direct inclusion in the concealment of this innovation, however something appeared to smother this evidently effective treatment to fix malignant growth. Overflowing accused conspiracy between the AMA and other clinical associations for excusing his logical cases. In any case, there doesn't give off an impression of being autonomous replication that his treatment worked.

Water-Powered Vehicles

As amazing as it would sound, there are really many working models for vehicles that can run on water. Obviously, you don't perceive any of these being gotten by the vehicle producers around the world.One of the most recorded (and celebrated) of these vehicles was a cart made by Stan Meyer. This amazing development accomplished a normal of 43 kilometers for every liter (100 mpg) of water. Partners near Meyer state that he was under a lot of coercion to offer the patent to his creation and end his examination into water cars.[9]But he would not be harassed into abandoning a working development that could totally change the world. Despite the fact that similar associates and companions yelled from the housetops that Meyer was harmed for his refusal to submit to enormous oil organizations, it is archived that Meyer kicked the bucket without notice from a cerebrum aneurysm.

Free Energy (Nikola Tesla)

Nikola Tesla was maybe one of the most notable innovators on the planet. In spite of the fact that everything made by his psyche probably won't be life changing, free power for the whole world ought to unquestionably liven up your ears. After effectively (and in very much reported cases) showing that he could remotely move power, Tesla made it realized that he was creating models that would enhance this wonder and force huge zones from a solitary tower.Though most at the time accepted this to be a genuine chance, Tesla's financing for the venture dwindled to nothing and his research facility with the model parts and plans strangely caught fire. This is maybe the most reported of all the smothered and smothered developments that you will never observe, and it is among the most all around huge.

Some other innovative inventions are :

INVENTION
YEAR
INVENTOR
COUNTRY
aerosol can
1926
Erik Rotheim
Norway
air conditioning
1902
Willis Haviland Carrier
US
airbag, automotive
1952
John Hetrick
US
airplane, engine-powered
1903
Wilbur & Orville Wright
US
airship

1852

Henri Giffard

France

alphabet

c. 1700â€"1500 BC

Semitic-speaking peoples

eastern coast of Mediterranean Sea

American Sign Language

1817

Thomas H. Gallaudet

US

animation, motion-picture

1906

J. Stuart Blackton

US

answering machine, telephone

1898

Valdemar Poulsen

Denmark

aspartame

1965

James Schlatter

US

aspirin

1897

Felix Hoffmann (Bayer)

Germany

assembly line

1913

Henry Ford

US

astrolabe

c. 2nd century

â€"

â€"

AstroTurf

1965

James M. Faria, Robert T. Wright

US
audiotape
1928
Fritz Pfleumer
Germany
automated teller machine (ATM)
1968
Don Wetzel
US
automobile
1889
Gottlieb Daimler
Germany
baby food, prepared
1927
Dorothy Gerber
US
bag, flat-bottomed paper
1870
Margaret Knight
US
Bakelite
1907
Leo Hendrik Baekeland
US
ball bearing
1794
Philip Vaughan
England
balloon, hot-air
1783
Joseph & Ã‰tienne Montgolfier
France
bandage, adhesive
1921
Earle Dickson
US
bar code

1952

Joseph Woodland

US

barbed wire

1874

Joseph Glidden

US

barometer

1643

Evangelista Torricelli

Italy

battery, electric storage

1800

Alessandro Volta

Italy

beer

before 6000 BC

Sumerians, Babylonians

Mesopotamia

bicycle

1818

Baron Karl de Drais de Sauerbrun

Germany

bifocal lens

1784

Benjamin Franklin

US

bikini

1946

Louis RÃ©ard

France

blood bank

late 1930s

Charles Richard Drew

US

blow-dryer

1920

Racine Universal Motor Co., Hamilton Beach Manufacturing Co.

US
bomb, atomic
1945
J. Robert Oppenheimer, et al.
US
bomb, thermonuclear (hydrogen)
1952
Edward Teller, et al.
US
boomerang
c. 15,000 years ago
Aboriginal peoples
Australia
Braille system
1824
Louis Braille
France
brassiere (bra)
1913
Mary Phelps Jacob
US
bread, sliced (bread-slicing machine)
1928
Otto Frederick Rohwedder
US
button
c. 700 BC
Greeks, Etruscans
Greece, Italy
buttonhole
13th century
â€"
Europe
calculator, electronic hand-held
1967
Jack S. Kilby
US
calculus

1680s

Sir Isaac Newton and Gottfried Wilhelm Leibniz (invented separately)

England and Germany (respectively)

calendar, modern (Gregorian)

1582

Pope Gregory XIII

Italy

camcorder

1982

Sony Corp.

Japan

camera, motion picture

1891

Thomas Alva Edison, William K.L. Dickson

US

camera, portable photographic

1888

George Eastman

US

can, metal beverage

1933

American Can Co.

US

can opener

1858

Ezra J. Warner

US

candle

c. 3000 BC

â€"

Egypt, Crete

canning, food

1809

Nicolas Appert

France

carbon-14 dating

1946

Willard F. Libby

US
cardboard, corrugated
1871
Albert Jones
US
cards, playing
c. 10th century
â€"
China
cash register
1879
James Ritty
US
cat litter
1947
Edward Lowe
US
catalog, mail-order
1872
Aaron Montgomery Ward
US
cellophane
1911
Jacques E. Brandenberger
Switzerland
celluloid
1869
John Wesley Hyatt
US
cement, portland
1824
Joseph Aspdin
England
cereal flakes, breakfast
1894
John Harvey Kellogg
US
chewing gum (modern)

c. 1870
Thomas Adams
US
chocolate
c. 3rd-10th century
Maya, Aztecs
Central America, Mexico
chronometer
1762
John Harrison
England
clock, pendulum
1656
Christiaan Huygens
The Netherlands
clock, quartz
1927
Warren A. Marrison
Canada/US
cloning, animal
1970
John B. Gurdon
UK
coffee, drip
1908
Melitta Bentz
Germany
coffee, decaffeinated
1905
Ludwig Roselius
Germany
coins
c. 650 BC
Lydians
Turkey
compact disc (CD)
1980
Philips Electronics, Sony Corp.

The Netherlands, Japan
compass, magnetic
c. 12th century
â€"
China, Europe
computed tomography (CT scan, CAT scan)
1972
Godfrey Hounsfield, Allan Cormack
UK, US
computer, electronic digital
1939
John V. Atanasoff, Clifford E. Berry
US
computer, laptop
1983
Radio Shack Corp.
US
computer, personal
1974
MITS (Micro Instrumentation Telemetry Systems)
US
concrete, reinforced
1867
Joseph Monier
France
condom, latex
c. 1930
â€"
â€"
contact lenses
1887
Adolf Fick
Germany
contraceptives, oral
early 1950s
Gregory Pincus, John Rock, Min Chueh Chang
US
corn, hybrid

1917
Donald F. Jones
US
correction fluid, white
1951
Bette Nesmith
US
cotton gin
1793
Eli Whitney
US
coupon, grocery
1894
Asa Candler
US
crayons, children's wax
1903
Edwin Binney, C. Harold Smith
US
cream separator (dairy processing)
1878
Carl Gustaf Patrik de Laval
Sweden
credit card
1950
Frank McNamara, Ralph Schneider (Diners' Club)
US
crossword puzzles
1913
Arthur Wynne
US
DDT
1874
Othmar Zeidler
Germany
defibrillator
1952
Paul M. Zoll

US
dentures
c. 700 BC
Etruscans
Italy
detector, metal
late 1920s
Gerhard Fisher
Germany/US
detector, home smoke
1969
Randolph Smith, Kenneth House
US
diamond, artificial
1955
General Electric Co.
US
diapers, disposable
1950
Marion Donovan
US
digital videodisc (DVD)
1995
consortium of international electronics companies
Japan, US, The Netherlands
dishwasher
1886
Josephine Cochrane
US
DNA fingerprinting
1984
Alec Jeffreys
UK
doughnut (ring) or donut
1847
Hanson Crockett Gregory
US
door, revolving

1888

Theophilus von Kannel

US

drinking fountain

c. 1905â€"1912

Luther Haws, Halsey W. Taylor (invented separately)

US

dry cleaning

1855

Jean Baptiste Jolly

France

dynamite

1867

Alfred Nobel

Sweden

elastic, fabric

c. 1830

Thomas Hancock

UK

electric chair

1888

Harold P. Brown, Arthur E. Kennelly

US

electrocardiogram (ECG, EKG)

1903

Willem Einthoven

The Netherlands

electroencephalogram (EEG)

1929

Hans Berger

Germany

electronic mail (e-mail)

1971

Ray Tomlinson

US

elevator, passenger

1852

Elisha Graves Otis

US
encyclopedia
c. 4th century BC or 77 AD

Speusippus (compliation of Plato's teachings) or Pliny the Elder (comprehensive work)
Greece or Rome
engine, internal-combustion
1859
Ã‰tienne Lenoir
France
engine, jet
1930
Sir Frank Whittle
UK
engine, liquid-fueled rocket
1926
Robert H. Goddard
US
engine, steam
1698
Thomas Savery
England
escalator
1891
Jesse W. Reno
US
eyeglasses
1280s
Salvino degli Armati or Alessandro di Spina
Italy
facsimile (fax)
1842
Alexander Bain
Scotland
fiber optics
1955
Narinder S. Kapany
India

fiberglass
1938
Owens Corning (corp.)
US
film, photographic
1884
George Eastman
US
flashlight, battery-operated portable
1899
Conrad Hubert
Russia/US
flask, vacuum (Thermos)
1892
Sir James Dewar
Scotland
food processor
1971
Pierre Verdon
France
foods, freeze-dried
1946
Earl W. Flosdorf
US
foods, frozen
c. 1924
Clarence Birdseye
US
Fresnel lens
1820
Augustin-Jean Fresnel
France
fuel cell
1839
William R. Grove
UK
genetic engineering
1973

Stanley N. Cohen, Herbert W. Boyer
US
Geiger counter
1908
Hans Geiger
Germany
glass
c. 2500 BC
Egyptians or Phoenicians
Egypt or Lebanon
glass, safety
1909
Ã‰douard BÃ©nÃ©dictus
France
greeting card, Christmas
1843
John Callcott Horsley
England
guillotine
1792
Joseph-Ignace Guillotin
France
guitar, electric
1941
Les Paul
US
gunpowder
c. 10th century
â€"
China or Arabia
hanger, wire coat
1903
Albert J. Parkhouse
US
helicopter
1939
Igor Sikorsky
Russia/US

holography
1948
Dennis Gabor
Hungary
hypodermic syringe
1853
Charles Gabriel Pravaz
France
in vitro fertilization (IVF), human
1978
Patrick Steptoe, Robert Edwards
UK
ink
c. 2500 BC
â€"
Egypt, China
insulin, extraction and preparation of
1921
Sir Frederick Grant Banting, Charles H. Best
Canada
integrated circuit
1958
Jack S. Kilby
US
Internet
1969
Advanced Research Projects Agency (ARPA) at the Dept. of Defense
US
iron, electric
1882
Henry W. Seely
US
irradiation, food
1905
â€"
US/UK
jeans
1873

Levi Strauss, Jacob Davis
US
JELL-O (gelatin dessert)
1897
Pearle B. Wait
US
jukebox
1889
Louis Glass
US
Kevlar
1965
Stephanie Kwolek
US
Kool-Aid (fruit drink mix)
1927
Edwin E. Perkins
US
laser
1958
Gordon Gould and Charles Hard Townes, Arthur L. Schawlow (invented separately)
US
laundromat
1934
J.F. Cantrell
US
lawn mower, gasoline-powered
c. 1940
Leonard Goodall
US
Lego
late 1940s
Ole Kirk Christiansen
Denmark
light bulb, incandescent
1879
Thomas Alva Edison

US
light bulb, fluorescent
1934
Arthur Compton
US
light-emitting diode (LED)
1962
Nick Holonyak, Jr.
US
linoleum
1860
Frederick Walton
UK
lipstick, tube
1915
Maurice Levy
US
liquid crystal display (LCD)
1963
George Heilmeier
US
lock and key
c. 2000 BC
Assyrians
Mesopotamia
locomotive
1829
George Stephenson
England
longbow
c. 1000
â€"
Wales
loudspeaker
1924
Chester W. Rice, Edward W. Kellogg
US
magnetic resonance imaging (MRI)

early 1970s
Raymond Damadian, Paul Lauterbur
US
margarine
1869
Hippolyte MÃ¨ge-MouriÃ¨s
France
matches, friction
1827
John Walker
England
metric system of measurement
1795
French Academy of Sciences
France
microphone
1878
David E. Hughes
UK/US
microscope, compound optical
c. 1600
Hans & Zacharias Jansen
The Netherlands
microscope, electron
1933
Ernst Ruska
Germany
microwave oven
1945
Percy L. Spencer
US
miniature golf
c. 1930
Garnet Carter
US
mirror, glass
c. 1200
Venetians

Italy
missile, guided
1942
Wernher von Braun
Germany
mobile home
1919
Glenn H. Curtiss
US
money, paper
late 900s
â€"
China
Monopoly (board game)
1934
Charles B. Darrow
US
Morse code
1838
Samuel F.B. Morse
US
motor, electric
1834
Thomas Davenport
US
motor, outboard
1907
Ole Evinrude
Norway/US
motorcycle
1885
Gottlieb Daimler, Wilhelm Maybach
Germany
mouse, computer
1963â€"64
Douglas Engelbart
US
Muzak

1922
George Owen Squier
US
nail, construction
c. 3300 BC
Sumerians
Mesopotamia
necktie
17th century
â€"
Croatia
neon lighting
1910
Georges Claude
France
nuclear reactor
1942
Enrico Fermi
US
nylon
1937
Wallace H. Carothers
US
oil lamp
1784
Aimé Argand
Switzerland
oil well
1859
Edwin Laurentine Drake
US
pacemaker, cardiac
1952
Paul M. Zoll
US
paper
c. 105
Ts'ai Lun

China
paper clip
1899
Johan Vaaler
Norway
paper towel
1931
Arthur Scott
US
parachute, modern
1797
André-Jacques Garnerin
France
parking meter
1932
Carl C. Magee
US
particle accelerator
1929
Sir John Douglas Cockcroft, Ernest Thomas Sinton Walton
Ireland/UK
pasteurization
1864
Louis Pasteur
France
pen, ballpoint
1938
Lazlo Biro
Hungary
pencil
1565
Conrad Gesner
Switzerland
periodic table
1871
Dmitry Ivanovich Mendeleyev
Russia
personal watercraft, motorized

1968
Bombardier, Inc.
Canada
petroleum jelly
1870s
Robert Chesebrough
US
phonograph
1877
Thomas Alva Edison
US
photocopying (xerography)
1937
Chester F. Carlson
US
photography
1837
Louis-Jacques-Mandé Daguerre
France
photography, instant
1947
Edwin Herbert Land
US
Play-Doh
1956
Noah W. & Joseph S. McVicker
US
plow, steel
1836
John Deere
US
pocket watch
c. 1500
Peter Henlein
Germany
polyethylene
1935
Eric Fawcett, Reginald Gibson

UK
polygraph (lie detector)
1921
John A. Larson
US
polyvinyl chloride (PVC)
1872
Eugen Baumann
Germany
Post-it Notes
mid-1970s
Arthur Fry (3M)
US
potato chips
1853
George Crum
US
printing press, movable type
c. 1450
Johannes Gutenberg
Germany
Prozac
1972
Ray W. Fuller, Bryan B. Molloy, David T. Wong
US
radar
c. 1904
Christian HÃ¼lsmeyer
Germany
radio
1896
Guglielmo Marconi
Italy
radio, car
early 1920s
William P. Lear
US
rayon

1884
Louis-Marie-Hilaire Bernigaud, count of Chardonnet
France
razor, electric
1928
Jacob Schick
US
razor, safety
c. 1900
King Camp Gillette
US
reaper, mechanical
1831
Cyrus Hall McCormick
US
record, long-playing (LP)
1948
Peter Carl Goldmark
US
refrigerator
1842
John Gorrie
US
remote control, television
1950
Robert Adler
US
respirator
c. 1955
Forrest M. Bird
US
revolver
1835â€"36
Samuel Colt
US
Richter scale
1935
Charles Francis Richter, Beno Gutenberg

US
rifle, assault
1944
Hugo Schmeisser
Germany
roller coaster
1884
LeMarcus A. Thompson
US
rubber, vulcanized
1839
Charles Goodyear
US
rubber band
1845
Stephen Perry
UK
saccharin
1879
Ira Remsen, Constantin Fahlberg
US, Germany
saddle (riding)
c. 200 BC
â€"
China
safety pin
1849
Walter Hunt
US
satellite, successful artificial earth
1957
Sergey Korolyov, et al.
USSR
satellite, communications
1960
John Robinson Pierce
US
saxophone

1846

Antoine-Joseph Sax

Belgium

Scotch tape

1930

Richard Drew (3M)

US

scuba gear

1943

Jacques Cousteau, Ã‰mile Gagnan

France

seat belt, automotive shoulder

1959

Nils Bohlin (Volvo)

Sweden

sewing machine

1841

BarthÃ©lemy Thimonnier

France

shoelaces

1790

â€"

England

silicone

1904

Frederic Stanley Kipping

UK

skateboard

1958

Bill & Mark Richards

US

skates, ice

1000 BC

â€"

Scandinavia

skates, roller

1760s

Joseph Merlin

Belgium
ski, snow
c. 2000â€"3000 BC
â€"
Sweden, Finland, Norway
skyscraper, steel-frame
1884
William Le Baron Jenney
US
slot machine
1890s
Charles Fey
US
snowmobile
1922
Joseph-Armand Bombardier
Canada
soap
600 BC
Phoenicians
Lebanon
soft drinks, carbonated
1772
Joseph Priestley
UK
sonar
1915
Paul Langevin
France
stamps, postage
1840
Sir Rowland Hill
UK
stapler
1866
George W. McGill
US
steamboat, successful

1807
Robert Fulton
US
steel, mass-production
1856
Henry Bessemer
UK
steel, stainless
1914
Harry Brearley
UK
stereo, personal
1979
Sony Corp.
Japan
stereophonic sound recording
1931
Alan Dower Blumlein
UK
stethoscope
1819
René-Théophile-Hyacinthe Laënnec
France
stock ticker
1867
Edward A. Calahan
US
stove, electric
1896
William Hadaway
US
stove, gas
1826
James Sharp
UK
straw, drinking
1888
Marvin Stone

US
submarine
1620
Cornelis Drebbel
The Netherlands
sunglasses
1752
James Ayscough
UK
sunscreen
1944
Benjamin Green
US
supermarket
1930
Michael Cullen
US
synthesizer, music
1955
Harry Olson, Herbert Belar
US
synthetic skin
1981
Ioannis V. Yannas, John F. Burke
US
tampon, cotton
1931
Earle Cleveland Haas
US
tank, military
1915
Admiralty Landships Committee
UK
tea bag
early 1900s
Thomas Sullivan
US
teddy bear

1902
Morris Michtom
US
Teflon
1938
Roy Plunkett
US
telegraph
1832â€"35
Samuel F.B. Morse
US
telephone, wired-line
1876
Alexander Graham Bell
Scotland/US
telephone, mobile
1946
Bell Laboratories
US
telescope, optical
1608
Hans Lippershey
The Netherlands
television
1923, 1927
Vladimir Kosma Zworykin, Philo Taylor Farnsworth
Russia/US, US
thermometer
1592
Galileo
Italy
thermostat
1830
Andrew Ure
UK
threshing machine
1778
Andrew Meikle

Scotland
tire, pneumatic
1888
John Boyd Dunlop
UK
tissue, disposable facial
1924
Kimberly-Clark Co.
US
tissue, toilet
1857
Joseph Gayetty
US
toaster, electric
1893
Crompton Co.
UK
toilet, flush
c. 1591
Sir John Harington
England
toothbrush
1498
â€"
China
tractor
1892
John Froehlich
US
traffic lights, automatic
1923
Garrett A. Morgan
US
transistor
1947
John Bardeen, Walter H. Brattain, William B. Shockley
US
typewriter

1868
Christopher Latham Sholes
US
ultrasound imaging, obstetric
1958
Ian Donald
UK
vaccination
1796
Edward Jenner
England
vacuum cleaner, electric
1901
Herbert Cecil Booth
UK
Velcro
1948
George de Mestral
Switzerland
vending machine
c. 100â€"200 BC
â€"
Egypt
Viagra
1997
Pfizer Inc.
US
video games
1972
Nolan Bushnell
US
videocassette recorder
1969
Sony Corp.
Japan
videotape
1950s
Charles Ginsburg

US
virtual reality

1989

Jaron Lanier
US
vision correction, laser
1987
Stephen Trokel
US
washing machine, electric
1907
Alva J. Fisher
US
wheel
about 3500 BC
proto-Aryan people or Sumerians
Russia/Kazakhstan or Mesopotamia
wheelbarrow
1st century BC
â€"
China
wheelchair
1590s
â€"
Spain
windmill
644
â€"
Persia
wine
before 4000 BC
â€"
Middle East
World Wide Web
1989
Tim Berners-Lee
UK
wristwatch, digital

1970

John M. Bergey

US

X-ray imaging

1895

Wilhelm Conrad RÃ¶ntgen

Germany

Zamboni (ice resurfacing machine)

1949

Frank J. Zamboni

US

zipper

1893

Whitcomb L. Judson

US

Citizen Scientist

From astronomy to zoology, wildlife surveys to exploring diaries from the First World War – if you have a passion or a curious mind there should be a project to suit you.

The principal behind citizen science is to use the power of collaborative volunteer research to explore or collect huge data sets. These are ones that researchers simply couldn't manage by themselves, and that computers aren't up to the task of analysing.

Take part and you could contribute to a genuine scientific breakthrough that changes our understanding of the world.

Who's it good for?

If you love the great outdoors there are lots of wildlife surveys to get involved with. Family friendly opportunities are available too, like worm and bug counts.

If you want to microvolunteer then there's a huge range projects you can dip into online as and when you want, for as little or as long as you fancy. So it's great if you are looking for something low-commitment, or maybe you have mobility issues. For housebound parents it's a quick something to try once the kids are in bed.

Where you can you do it?

Wildlife surveys tend to be done outdoors, or at least looking out of your kitchen window. Many of the projects are online though and you can do them at home, in the office, on holiday... anywhere you have an internet connection.

What does it involve?

It depends on the project, but there are two main sorts. Either observing the world, recording what you see and submitting the data. Or going through pre-existing data – such as star photos – and logging what you find.

Many projects require no specialist knowledge or skills – anything you need to know is explained when you take part. Others draw on the knowledge of amateur gardeners, bird spotters, naturalists, star gazers...

Who does it help?

There are a huge range of subjects for citizen science projects including astronomy, photo-tagging, wildlife surveys, public health, air pollution, weather... What they all do is use data sets to help understand the world around us.

Commitment Level?

Many of the online actions are one off – so you can dip into citizen science as and when you have a moment. Some surveys focus on the natural world in a brief burst – called a bioblitz – and others track it for longer periods. It's possible to choose a project that suits the level of commitment you are looking for.

Getting Started

Zooniverse has a wide range of online citizen science projects – from hunting for comets, to tagging penguins, to exploring the secret lives of artists through their notebooks. And microvolunteering site Help from Home has a section specifically devoted to citizen science opportunities

There are a huge number of nature related surveys to get involved with. Some are ongoing, such as the British Trust for Ornithology's Garden BirdWatch and The

Conclusion

There are several things which shows impactful in ancient technologies. So in future science the main focus is impactful frugal innovation for making invention stronger strengther and resilience to society.

Electronic Cloth

Physicists at Wake Forest University have developed a fabric that doubles as a spare outlet. When used to line your shirt — or even your pillowcase or office chair — it converts subtle differences in temperature across the span of the clothing (say, from your cuff to your armpit) into electricity. And because the different parts of your shirt can vary by about 10 degrees, you could power up your MP3 player just by sitting still. According to the fabric's creator, David Carroll, a cellphone case lined with the material could boost the phone's battery charge by 10 to 15 percent over eight hours, using the heat absorbed from your pants pocket.

The New Coffee

Soon, coffee isn't going to taste like coffee — at least not the dark, ashy roasts we drink today. Big producers want uniform taste, and a dark roast makes that easy: it evens out flavors and masks flaws. But now the best beans are increasingly being set aside and shipped in vacuum-sealed packs (instead of burlap bags). Improvements like these have allowed roasters to make coffee that tastes like Seville oranges or toasted almonds or berries, and that sense of experimentation is trickling down to the mass market; Starbucks, for instance, now has a Blonde Roast. As quality continues to improve, coffee will lighten, and dark roasts may just become a relic of the past. Oliver Strand

Monitoring Multi Tasker

Your spandex can now subtly nag you to work out. A Finnish company, Myontec, recently began marketing underwear embedded with electromyographic sensors that tell you how hard you're working your quadriceps, hamstring and gluteus muscles. It then sends that data to a computer for analysis. Although the skintight shorts are being marketed to athletes and coaches, they could be useful for the deskbound. The hope, according to Arto Pesola, who is working on an advanced version of the sensors, is that when you see data telling you just how inert you really are, you'll be inspired to lead a less sedentary life. Gretchen Reynolds

Clean Hair No hands

This 15-minute shampoo treatment begins when you lean your head back into a machine that looks like a sink at the salon. First it maps your scalp, then it shoots streams of warm water and foam shampoo from its 28 nozzles before 24 silicone "fingers" work up a lather. One conditioning mist, scalp massage and light blow-dry later, you're don

The first is permanent sunblock. No one likes putting the stuff on, so there should be a one-time treatment that embeds the skin with a permanent level of S.P.F. 30, akin to having Lasik eye surgery once and then forgetting about it. Sunburn vanquished like smallpox. The other is the "brain map" — a technology that maps out every neural connection in your mind and then, effectively, stores your brain on your hard drive. That information — more than your DNA even — is you.

Traffic System

Traffic jams can form out of the simplest things. One driver gets too close to another and has to brake, as does the driver behind, as does the driver behind him — pretty soon, the first driver has sent a stop-and-go shock wave down the highway. One driving-simulator study found that nearly half the time one vehicle passed another, the lead vehicle had a faster average speed. All this leads to highway turbulence, which is why many traffic modelers see adaptive cruise control (A.C.C.) — which automatically maintains a set distance behind a car and the vehicle in front of it — as the key to congestion relief. Simulations have found that if some 20 percent of vehicles on a highway were equipped with advanced A.C.C., certain jams could be avoided simply through harmonizing speeds and smoothing driver reactions. One study shows that even a highway that is running at peak capacity has only 4.5 percent of its surface area occupied. More sophisticated adaptive cruse control systems could presumably fit more cars on the road. Tom Vanderbilt

When a quarter of the vehicles on a simulated highway had A.C.C., cumulative travel time dropped by 37.5 percent.

In another simulation, giving at least a quarter of the cars A.C.C. cut traffic delays by up to 20 percent.

By 2017, an estimated 6.9 million cars each year will come with A.C.C.

Anti-slouch Screen

If you slump down when you're typing on an ErgoSensor monitor by Philips, it'll suggest that you sit up straighter. To help office workers avoid achy backs and tired eyes, the device's built-in camera follows the position of your pupils to determine how you are sitting. Are you too close? Is your neck tilted too much? Algorithms crunch the raw data from the sensor and tell you how to adjust your body to achieve ergonomic correctness. The monitor can also inform you that it's time to stand up and take a break, and it will automatically power down when it senses that you've left.

A glass that turns water into wine

Vocktail, created by scientists from Singapore, is a glass that can change the flavor and color of the drink inside it. This glass with a "virtual cocktail" is connected to a mobile app that allows you to control the settings for the liquid.

18. "Smart" glasses

There are many cool things about Vue glasses: they can make calls, transmit music, and be a navigator, pedometer, and calorie counter. The main thing is that they have a "Find my glasses" function, so you won't need to spend hours looking for your glasses. Vue glasses look like regular glasses (there's also an option for sunglasses), they are controlled by touch, and they have a special case with wireless charging.

A toothbrush that cleans your teeth by itself

Amabrush brushes your teeth much more thoroughly than a regular brush, and it does its job in only 10 seconds. You just have to put it in your mouth and connect the device to your smartphone via Bluetooth.

A pendant that turns speech into text

Senstone can be attached to clothes or hung on one's neck as a pendant. With just one click, the device starts transforming speech into text with 97% precision. It's the dream of all students: now you don't have to write down lecture notes! The pendant "understands" 12 languages.

A mat with a built-in alarm clock

Ruggie is for those who can't wake up with a regular alarm clock. To deactivate the alarm, you have to step on it with both feet for 3 seconds. This is enough time for your brain to cope with the horror of awakening and starting a new day.

A pancake printer

PancakeBot can bake a pancake in any shape, from a flower up to your favorite cartoon character. Kids can draw their future pancakes and then watch the device printing their breakfast. Here you can see a hypnotizing video about the device.

A water-sifting straw

The LifeStraw channel evacuates 99.9% of microscopic organisms and 96.2% of infections. It was at first made for those in crisis conditions and for individuals living in creating nations without enough new clean water. Notwithstanding, this gadget turned out to be exceptionally well known among world explorers. You can utilize this gadget to drink water from a stream or a lake. It very well may be valuable when you don't know of the nature of the water you are drinking.

Bundling that changes shading if the item inside is terminated

An organization named Braskem worked together with American and Brazilian researchers to make a sort of plastic that can change its shading relying upon pH levels. It tends to be utilized to create bundling for short-lived nourishments. Before long we will see with our own eyes how new the milk in a market truly is.

A windshield show

Carloudy trades route data with your cell phone or some other Bluetooth gadget. The picture is transmitted legitimately onto the windshield without diverting the driver. The gadget is constrained by voice orders, which permits a driver to keep their hands on the controlling wheel.

A "brilliant" coat

Brilliant Parka is a fantasy for the individuals who continually lose their gloves since they are incorporated into the "savvy" cover alongside a scarf and top. The length of the coat can be modified, and there are various contraption pockets inside. There is likewise a geotag inside, which can be valuable if the coat is lost or taken. Obviously, it's warm as it was extraordinarily made for cold Canadian winters.

A pocket film

This gadget that seems as though a soft drink can is a projector and remote amplifier. You can have your film any place you need: on your room roof or on a divider. Container's battery runs for 2.5 hours in film mode and 40 hours in amplifier mode.

A stain-repellant shirt

Fooxmet is a shirt produced using a hydrophobic cotton material. It lets air through, yet it repulses any fluids – from water to ketchup. The primary concern is you don't have to press it as it barely ever gets wrinkled.

A pickpocket-verification knapsack

LocTote seems as though a customary knapsack, yet it's really a delicate vault for your stuff. It can't be cut or set ablaze. It must be opened utilizing a lock, and you can even leave it unattended. It's a helpful thing for voyaging that will disillusion all the pickpockets of the world.

An electronic baggage tag

Mu Tag is a little electronic label that is connected to things you would prefer not to lose: a sack, keys, a pooch neckline, an envelope with records, and so on. If there should be an occurrence of misfortune or burglary, this little thingamabob can assist with following the position by means of an application.

Antistress block

Squirm Cube was created for the individuals who like gnawing a pen or destroying paper while they think. It's a little 3D shape with a joystick, a mix lock, a turning ball, and a delicate touch surface.

An ear cleaner that leaves no way for earwax

EarScope is for clean oddities. It was designed to let you see inside your ears and clean them with extreme accuracy. With brilliant LED lights, EarScope can likewise be utilized for peering inside dull and difficult to-arrive at places.

The primary E-Ink tattoo machine

moodINQ was created to upset the universe of tattoos. It is utilized with an exceptional "canvas" that must be embedded in your skin. At that point you can transfer a structure to it. You can change the structure or delete it at whatever point you need. Presently you can even keep your basic food item list on your arm!

A touchless sack sealer

This small contraption is made for the individuals who like purchasing contributes colossal cumbersome sacks. It is called iTouchless, and it will assist you with keeping every one of your snacks pleasant and crunchy by making an impenetrable seal in a matter of moments

References

- https://www.ancient-origins.net/artifacts-ancient-technology/6-advanced-ancient-inventions-beyond-modern-understanding-002652
- http://www.ancientpages.com/2019/07/23/10-remarkable-advanced-ancient-technologies-ahead-of-their-times/
- https://www.theclever.com/15-pieces-of-ancient-technology-we-still-dont-understand/
- https://www.aquiziam.com/ancient-technology/
- https://www.learning-mind.com/nikola-teslas-innovations/
- https://www.scientificamerican.com/article/follow-up-what-is-the-zer/
- https://listverse.com/2009/07/31/top-10-amazing-lost-or-suppressed-inventions/
- https://factspage.blogspot.com/2013/06/perpetual-motion-machine.html
- https://www.greenoptimistic.com/karpen-pile/
- https://listverse.com/2017/06/08/10-great-inventions-we-will-probably-never-see/
- https://www.livescience.com/55944-perpetual-motion-machines.html
- https://teslagenerator.com/tesla-generator-perpetual-energy-for-free/
- https://theplaidzebra.com/zero-point-energy-generators/
- https://talentculture.com/innovation-and-technology-the-irreplaceable-human-factor/
- https://engineering.eckovation.com/51-greatest-innovation-time/
- http://www.arthistoryarchive.com/arthistory/architecture/Ancient-Architecture.html
- https://www.thoughtco.com/architecture-timeline-historic-periods-styles-175996
- https://www.planndesign.com/articles/2750-how-different-ancient-and-modern-architecture
- https://www.umdaschgroup-ventures.com/en/magazine/frugal-innovations-for-a-sustainable-growth
- https://www.businessmagazinegainesville.com/frugal-innovation/
- https://blogs.sap.com/2020/06/14/frugal-innovation-how-to-find-opportunity-in-a-storm-of-adversity/

- https://thrive.dxc.technology/2020/03/05/embracing-reverse-and-frugal-innovation-in-emerging-markets/
- http://eadi-nordic2017.org/2016/08/30/globalization-of-the-future-how-can-frugal-innovation-foster-economic-social-and-environmental-sustainability-working-group-frugal-innovation-and-development/
- https://en.wikipedia.org/wiki/Zero-point_energy
- http://blog.hasslberger.com/2013/11/german_inventor_solves_permane.html
- https://brightside.me/wonder-curiosities/19-cool-inventions-for-those-who-want-to-live-in-the-future-now-405360/
- https://www.bbc.co.uk/programmes/articles/4BZZdHm64S051q2lnZ1Nr7p/citizen-science

About Author

Professor Doctor Sanjay Rout

Prof. (Dr.) Sanjay Kumar Rout is an International Researcher, Innovator, Speaker, Author, Journalist and Policy Expert, Coach. He is well known and highly respective dignitary in the field of Research Development & Innovation work in major domain of Development Management, Policy Research, Public Policy, Business, Economics, Finance, Law, Social Science, Education, Technology and other Fields.

He is Global Scientist (NCCHWO). Prof. (Dr.) Sanjay Kumar Rout has been distinguished Researcher, Startup Mentor Innovator, who consistently demonstrates his research work excellence in field of Research & development, Innovations with greater efficiency, productivity, and quality Innovations & research models.

Health, Governance, Technology, Business Management & Academics.

He had received many National / International Fellowship & Awards in several categories for his eminent work in Innovation, Management, Research, Sustainability, and Social Development. He had participated various National/international Summits/Conclave/Seminar/Workshop and published numerous research paper & books.

For his work he had been Honored by many organization as :

- World Top Futher Thought Leader in Open innovation & Business
- National Innovator Award
- Out Standing Researcher Award

- Best Young Scientist Award
- Best Speaker Award
- World Top 50 Futher Thought Leader in Data Privacy & Agile
- Best Global Scientist, Policy cum Journalist Award

His academic credentials contain different achievements from renowned university /institutions like—NIT, IIM, IIT, University of Pennsylvanian, and University of Washington, Imperial College London, John Hopkins University & others. Including Several achievement's, he holds three Ph.D.& one D.Sc (Higher Doctorate) as in his research career.

He is an global certified professional from international acclaimed organization like Google,WHO, BCG,World Bank, Amazon,UNICEF, SAS,UN, European Union, IBM, Asian Development Bank, FAO, Cisco, IRCC,GoI,UNDP & others. And he had worked for various global projects in multiple thematic areas.

www.ingramcontent.com/pod-product-compliance
Lightning Source LLC
Chambersburg PA
CBHW061508180526
45171CB00001B/92